Green Energy and Technology

For further volumes:
http://www.springer.com/series/8059

Ray Galvin · Minna Sunikka-Blank

A Critical Appraisal of Germany's Thermal Retrofit Policy

Turning Down the Heat

Ray Galvin
Minna Sunikka-Blank
Department of Architecture
University of Cambridge
Cambridge
UK

ISSN 1865-3529   ISSN 1865-3537 (electronic)
ISBN 978-1-4471-7018-1   ISBN 978-1-4471-5367-2 (eBook)
DOI 10.1007/978-1-4471-5367-2
Springer London Heidelberg New York Dordrecht

Printed on acid-free paper

Springer is part of Springer Science+Business Media (www.springer.com)

Contents

Abbreviations

BMVBS	(Bundesministerium für Verkehr, Bau und Stadtentwicklung): Federal Ministry for Transport, Building and Urban Development, shortened in the text to 'Housing Ministry'
BMU	(Bundesministerium für Umwelt, Naturschutz und Reaktorsicherheit): Federal Ministry for the Environment, Conservation and Nuclear Reactor Safety
BMWi	(Bundesministerium für Wirtschaft und Technologie): Federal Ministry of the Economy and Technology
DENA	(Deutsche Energie-Agentur): German Energy Agency
DIN	(Deutsches Institut für Normung): German Institute of Standards
EnEV	(Energieeinsparverordnung): Energy saving regulations (for buildings)
EnEG	(Energieeinsparungsgesetz): Energy saving law (for buildings)
H_T	Heat transmission losses, i.e., the rate of heat energy loss through the building envelope as a whole, expressed in W/m^2K
IWU	(Institut Wohnen und Umwelt): Institute for Housing and Environment
KfW	(Kreditanstalt für Wiederaufbau): German Development Bank
kWh/m^2a	kilowatt-hours per square metre per year, a measure of the quantity of energy consumed in a building
Living area	(Wohnfläche): The floor area of a dwelling inside the front door of the apartment itself, not counting other areas within the building (e.g., basement, attic, stair wells, landings)
Q_M	The actual, annual measured quantity of primary energy consumed for space and water heating, expressed in kWh/m^2a
Q_P	The theoretical annual quantity of primary energy required to keep a dwelling at a standard temperature continuously and to supply water heating, expressed in kWh/m^2a
U-values	Transmission losses through specific components of the building envelope, expressed in W/m^2K
UBA	(Unweltbundesamt): Federal Environment Bureau

Useable area   (Nutzfläche): The floor area of a dwelling including its 'living' area (the area inside its front door) plus a relevant proportion of the area of the stairways, landings, basement, attic and other areas inside the building, used in Germany as the floor area on which calculations of the dwelling's heating energy consumption are based

W/m^2K Watts of heat transfer through a medium, per square metre of the medium, per degree Kelvin difference in temperature between its two sides

Wirtschaftlich   Economically viable

WSVO   (Wärmeschutzverordnung): Heat retention regulations (for buildings)

About the Authors

Dr. Ray Galvin has an interdisciplinary background in engineering, philosophy social psychology, climate change science and environmental policy studies. In recent research he has sought to bring theoretical approaches from the physical and social sciences together, to develop transdisciplinary frameworks for environmental science research. Over the past four years his research focus has been on thermal retrofits in Germany, investigating how government policy, the physical properties of buildings and the motivation and behaviour of households interplay to inhibit or enable fuel savings. His research has been conducted through the Universities of East Anglia and Cambridge in the UK and RWTH-Aachen in Germany. He also has 30 years' practical experience in managing a portfolio of rental housing.

Dr. Minna Sunikka-Blank is a Senior Lecturer at the Department of Architecture in Cambridge University and a Fellow and Director of Studies in Architecture in Churchill College. Her research focuses on comparative policy analysis and she has published several books and research papers on the subject. She is particularly interested in how national governments can improve their sustainable building policies so as to increase feasible, cost-efficient and legitimate carbon reductions in the housing stock, supported by household behavioural change. She is a registered architect and has worked on urban renewal projects and sustainable retrofits in Finland and in The Netherlands.

Chapter 1
Introduction

1.1 The German Case

This book focuses on thermal retrofits of homes in Germany, but its findings have a much wider application than that one country. Germany is one of a number of northwest European countries that have taken major strides in recent decades to reduce domestic heating fuel consumption: not only by continually tightening the thermal standards for new buildings, but also by setting compulsory thermal standards for existing homes when they are being repaired or renovated. Most European countries have compulsory thermal standards for the replacement of windows, doors, roofs, and boilers, but in 2002 Germany took the further step of extending this to cover the entire outer surface area of the building. If any substantial repairs or maintenance are being done on the outer walls, basement ceiling, or loft, these must include insulation to specified standards. Even in the absence of repairs, any attempt to insulate these areas must reach these standards.

These rules are set down in the 'Energy Saving Regulations' (*Energieeinsparverordnung*), known as the 'EnEV'. The introduction of mandatory standards for existing housing was possible because the technology for thermally upgrading buildings is well advanced in Germany, and also well distributed. Over the past 10 years the German Federal government has vigorously promoted thermal retrofitting of existing homes, and results are to be seen throughout the country. The German Energy Agency (DENA) has monitored comprehensive thermal retrofits of residential buildings of every type and age, demonstrating that the heating energy requirements of almost any building can be reduced by 60% or more. The Federal Housing Ministry (*Bundesministerium für Verkehr, Bau und Stadtentwicklung*—BMVBS) offers a suite of subsidies for thermal retrofit projects that are designed to do substantially better than the standards set down in the EnEV, and the Federal Ministry of Economics and Technology (*Bundesministerium für Wirtschaft und Technologie*—BMWi) finances energy advisors to assist homeowners toward retrofit decisions that conform to EnEV standards or better. Further incentives and assistance are offered by *Länder* (states) and municipalities, so that there are over 1,000 programs, in different geographical areas and at various

R. Galvin and M. Sunikka-Blank, *A Critical Appraisal of Germany's Thermal Retrofit Policy*, Green Energy and Technology, DOI: 10.1007/978-1-4471-5367-2_1, © Springer-Verlag London 2013

levels, for homeowners to turn to for help in planning or carrying out a thermal retrofit. The policy discussion in this book focuses on Federal initiatives, as these set very strict parameters for thermal standards, within which other authorities have to work.

Germany's thermal retrofit policy can be seen in the context of German and EU climate and energy policy. In Europe around one-third of all energy consumption takes place in buildings. More than half of this happens in homes, and 80–90% of that is for space and water heating. This means that between 12 and 20% of the energy consumed in European countries goes on domestic heating. In Germany the figure is around 14.6%. This makes domestic heating a major issue: energy is expensive; there is concern about security of supply; and domestic heating fuels such as oil, coal, natural gas, and electricity from fossil fuel burning plants produce greenhouse gas (GHG) emissions. The European Union has a commitment to reduce GHG emissions by 20% by 2020, and Germany has adopted the goal of an 80% reduction by 2050, a target that also applies to domestic heating consumption.

1.2 Success or a Stalled Project?

For other countries considering adopting a policy similar to that of Germany, an essential question is how likely Germany's retrofit policy is to achieve the 80% reduction by 2050. Our analysis indicates that despite these ambitious efforts and encouraging signs, thermal retrofits in Germany are coming nowhere near the goals of energy and CO_2 emission reduction the government has set. Since most domestic heating fuel in Germany is oil and gas burnt on site, GHG reductions are almost directly proportional to fuel consumption reductions, so we can use roughly similar percentage reduction figures for each. Hence, to reach the 80% goal would require 2.6% of the building stock to be retrofitted every year to an average depth of 80%, or total annual reductions in heating energy consumption of 2.1% per year. But the achievement of the last 10 years has been 1% of the building stock retrofitted per year to an average depth of 25%, meaning a total annual reduction of 0.25%. Less than one-eighth the required reductions are currently being achieved.

This book investigates why this is so, but with the positive aim of exploring how Germany could come closer to achieving its goal and increase the retrofit rate. There are important lessons to be learnt from the German experience of thermal upgrades of existing homes. In this book, we look in close detail at Germany's thermal retrofit experience from a number of complementary points of view: technical; economic; policy-based; and the somewhat unpredictable factor of human behavior within the homes that are, or are not (yet), thermally retrofitted. Only through an interdisciplinary examination, which understands the physical realities and diversity of the existing buildings, together with the issues faced by occupants, can ways be found to increase the annual rate of retrofits and energy saved.

Through this process we develop a suggested policy program which, we believe, could increase the rate of energy savings from its current 0.25% to around 1.2%—not enough to reach the 80% policy goal, but nearly five times as high as at present. The program we will suggest in Chap. 9 is based on a three-pronged policy approach that can be summarized under the acronym CUT: a combination of Cost-effective retrofits; User behavioral change; and Top-end measures. Currently almost all the focus in Germany is on the latter. However, there is great fuel saving potential in the middle strand, user behavior, as pointed out also by the EU Commission study on behavioral climate change mitigation (Faber and Schroten 2012). Although our scheme arose out of an exacting study of the German thermal retrofit experience, we think it could have a wider application throughout Europe, if national characteristics of the building stock, tenure, and policy structure are taken into account.

1.3 Motivation for this Research

Our previous research has investigated the interplays between government policy on thermal retrofits and the physical characteristics of the buildings at which the policy was aimed, focused on Germany, the Netherlands, and the UK. This book was initiated by our common identification of serious discrepancies between the policy and the characteristics of the building stock (see e.g., Galvin 2010, 2011, 2012, 2013; Sunikka 2001, 2003, 2006a; Sunikka and Boon 2004; Sunikka-Blank et al. 2012). Buildings are obstinate objects that slavishly obey the laws of physics with respect to each structure's physical peculiarities. When thermal upgrades are done to them according to official ordinances, they will not necessarily save the quantities of energy prescribed in policy documents. Further, the people living in the buildings have their own set of rules and responses, which often render the calculations of energy experts irrelevant to what happens in practice.

These findings led us to become intensely interested in household heating behavior. There are important developments in recent scholarship on consumer behavior and its effects on energy consumption. This utilizes social science for investigating the behavior itself, but tends to leave the cutting-edge study of the physics of buildings, and often also of policy, to experts in those fields (e.g., Hargreaves et al. 2010; Shove et al. 2008). Building on our interdisciplinary background of architecture, energy policy studies, engineering, social psychology, and environmental science, our approach in this book is to add a scholarly approach to consumer behavior to the two foci of study—policy and the physical properties of buildings—we were already conversant with.

To lay a foundation for this approach we have undertaken a number of inter-related research projects designed in part to 'flush out' behavioral factors that may be contributing to outcomes in the German experience of thermal retrofits of existing homes. We use standard methodologies in policy studies and physics (sometimes also micro-economics) that show up where the straightforward application of the upgrade measures demanded by policy is thwarted, not just by the obstinate physics of the buildings, but also by factors such as behavior, that cannot be explained by mere physics. These can be identified as areas where social science studies in consumer behavior are needed. Hence, much of this book points to the need for social science-based studies in certain areas, but within a context clearly defined by what we know of policy and of the material characteristics of the buildings in which the people are doing their consumption. An example of such a social science study—which goes beyond the scope of this book but which illustrates the type of study it could lead to—is that on energy-inefficient home ventilation practices (see Galvin 2013).

1.4 Fieldwork

The fieldwork that contributed to the knowledge and understanding presented in this book took place over the last four years. This included formal, recorded interviews with key policy actors involved in developing and promoting Germany's thermal retrofit policy: Federal MPs and their parliamentary research assistants; experts from the Housing Ministry (*Bundesministerium für Verkehr, Bau und Stadtentwicklung*—BMVBS), the Federal Planning Bureau (*Bundesamt für Bauwesen und Raumordnung*—BBR), and the German Energy Agency (*Deutsche Energie-Agentur*—DENA); State and municipal politicians and civil servants in North Rhine-Westphalia, Hamburg, Munich, Augsburg, Cologne, and Düsseldorf; Federal expert advisors from the Institut Wohnen und Umwelt (IWU), the Passive House Institute, the Fraunhofer Institute, and the Technical University of Munich; research staff of the German Association of Housing Providers (GdW); and leading practitioners in the application of the EnEV standards to large retrofit projects (21 recorded interviews; see Galvin 2011, 2012). Non-recorded, follow-up interviews, and discussions were held with key Federal MPs and relevant staff of key NGOs in Berlin more recently, to gauge responses to changing perceptions of the efficacy of the EnEV among practitioners. These encounters were in the context of hundreds of hours of on-site observations of buildings prior to, during, after and in the absence of thermal retrofitting, and dozens of discussions with local, smaller scale practitioners.

In addition, interviews were conducted with a selection of owner-occupiers throughout Germany. This was partly to hear their views on how the retrofit regulations impacted on their own situation, but also to collect concrete examples of thermal retrofit experiences, which could be (anonymously) related to policy actors to hear their responses. Interviewees were male and female, aged 30–75, in the cities of Augsburg, Baden, Berlin, Cottbus, Erfurt, Lübeck, Lüneburg, and Würzburg, plus villagers in rural regions of northern Bavaria. Their homes included detached and semi-detached houses, and apartments in multi-storey blocks. Of the 20 formally interviewed, 14 permitted the interviews to be recorded and transcribed. These encounters were supplemented by over 100 further discussions with renters and owner-occupiers throughout Germany over a period of four years (Galvin 2011), plus direct observations of many homes and detailed calculated case studies of thermal upgrade effects on a number of these (Galvin 2010).

This has been further supplemented by document research, including, in particular, over 100 reports, from German research institutes, on research into various aspects of domestic heating energy consumption and thermal retrofitting. Relevant documents among these are specified in each chapter and named in detail in the references.

Beyond Germany our fieldwork has included comparative analysis of retrofit policies in European countries (e.g. Sunikka 2006a) and the EU, and the effectiveness of the European Energy Performance of Buildings Directive (EPBD) in particular (Beerepoot and Sunikka 2005). This has included an overview of building statistics in the EU countries (Meijer et al. 2009) as well as country-specific case studies on sustainable housing management and retrofits, for example in the Netherlands (Sunikka and Boon 2002, 2003, 2004; Sunikka 2006b) and the UK (Sunikka-Blank et al. 2012), and a case study in Aachen, Germany, including a year-long test on the relative contributions of thermal properties and user behavior to heating fuel consumption. When these energy-efficient retrofits have been implemented in practice, two factors have emerged as main barriers for a wider application: long payback times of retrofit measures, that somehow always seem to exceed the optimistic policy estimates, and economically non-rational user behavior, that distorts the expected energy savings and payback times. This book offered us an opportunity to explore how these practical barriers could be overcome in a policy.

1.5 Themes of the Book

Chapter 2 gives an overview of German Federal policy on thermal retrofits of existing homes. It shows how this is integrated with Germany's climate policy, which sets for it the demanding target of 80% reductions in home heating energy consumption by 2050. The main policy instruments for achieving this goal are: regulations; a promotional program based largely on the claim that retrofits to Federal standards always pay back; a set of subsidies for retrofits designed to do better than the regulations; and subsidized on-site visits of energy advisors. The chapter also introduces the evidence that annual nationwide savings achieved through thermal retrofits amount to around 0.25%.

Chapter 3 sets the German program in the wider context of Europe-wide issues and EU policy, which aims for 20% reductions in GHG emissions by 2020, together with 20% of energy produced from renewable sources and a 20% increase in energy efficiency. It contrasts Germany's compulsory thermal upgrade approach with voluntary, market-led approaches in other countries, and compares the effectiveness of different country's subsidy systems. It also brings into focus the general issue of how household behavior impacts on home heating energy consumption levels.

Chapter 4 examines the technical potential of thermal upgrades in Germany's housing stock. While economics often proscribes what is practically feasible, this chapter focuses on the technical potential and the challenges of bringing German dwellings up to regulation thermal standards. The dictates of physics, geometry, and the material peculiarities of the buildings that actually exist in Germany offer challenges to the stiff demands of the regulations. Further, measuring the thermal performance of buildings pre- and post-retrofit can only be done by making uniform assumptions about user behavior, since it is occupants who control the day-to-day heating and ventilation of their homes. Hence, even in the apparently purely technical domain of designing thermal upgrades, human behavior adds a further factor that makes final outcomes unpredictable.

Chapter 5 makes a systematic attempt to quantify and find possible explanatory factors for a phenomenon that has recently become a source of concern to German scholars: the discrepancies between theoretical, calculated heating energy consumption, and actual measured consumption. On average, actual consumption is 30% below theoretical consumption. This percentage gap increases as theoretical consumption increases, rising to 60% or more, but for low energy homes it goes into reverse. We label this phenomenon the 'prebound effect' (in contrast to the well-known 'rebound effect'), as it is evident in dwellings prior to retrofitting. In part this phenomenon may be due to flaws in the methodology for calculating theoretical consumption, but on closer examination we find that much of it cannot be explained by technical matters. We suggest this points to the need for social science study of household behavior, to find whether human factors are responsible for the gap between calculated and measured heating energy consumption, particularly in older, thermally inefficient homes. If human factors are at work, we need to know to what extent, in what ways, and why. The prebound effect also shows that the common German practice of using calculated pre- and post-retrofit consumption figures to estimate the amount of energy that is saved through retrofitting greatly overestimates the savings that are being achieved.

In Chap. 6 we examine the economics of thermal retrofits to EnEV standards. German law limits thermal regulations to setting upgrade targets no higher than that which pays back, through fuel savings, over the technical lifetime of the upgrade measures. Having set the regulations to what they believe this level is in the current market, policymakers promote thermal retrofitting on the basis that it pays for itself. We investigate a number of controversial points in the cost-benefit models policymakers use to prove that thermal retrofits are 'economically viable'. These include: the technical limitations discussed in Chap. 4; the way the accounting is done; and the reduction in actual fuel savings due to the prebound effect. Our calculations—based on case studies of our own and others—do not support the view that these 'top-end' thermal retrofits are economically viable. Ironically, there are more modest thermal upgrade measures that are far more likely to pay back, but these are often illegal in Germany as they do not meet the standards of the EnEV. Nevertheless, there may be other benefits in top-end retrofits even if they do not pay back, and a higher retrofit rate might ensue from a clearer policy focus on these other benefits, while also allowing and fostering more modest retrofit measures.

Chapter 7 investigates a phenomenon that has so far received little attention in studies on German (or other) thermal upgrades. Over the past decade, the actual, measured consumption of domestic heating energy has been falling in Germany by around 1.85% per year, but this fall is too big to be explained by thermal retrofits and replacement of old stock with new, energy-efficient buildings. Based on statistics gleaned from a number of recent studies, we set out a method for disaggregating the factors contributing to this fall in consumption: new builds; thermal retrofits; and 'unexplained' or 'non-technical' factors. We find that over 80% of the reduction cannot be explained by the first two categories, and suggest the bulk of the savings must therefore be due to demographic or user behavioral influences. Again, this points to the need for a social science investigation to find out what has been happening in non-retrofitting households whose consumption has fallen so significantly in recent years. Also, if some such households have been consciously reducing their heating demand, we need to know if this is a response to increased fuel prices, growing environmental awareness, or other yet-unknown factors.

Chapter 8 is a mathematical modeling project that brings together these suggested user behavior factors with the economics and technical features of thermal retrofits. We begin with the basic form of the cost-benefit equations generally used to estimate economic viability of thermal upgrades in Germany and other countries. We ask what would happen to these equations if a factor for price elasticity of demand for domestic heating fuel were integrated into them: how would it affect economic viability, payback time, and the quantity of energy and CO_2 saved as a result of the retrofit measures? Then, building on existing empirical studies of household response to fuel price increases, together with our own findings in Chap. 7, we calculate a percentage value for price elasticity and plug it into the new cost-benefit equations we have derived. This shows that economic viability is less certain, payback takes longer, and fuel and CO_2 savings are lower, than is calculated by the standard models. This, too, suggests the need for a social science study of user behavior: if modeling shows that price-sensitive user behavior would distort the modeled results so significantly, we need urgently to know whether or not, and to what extent, this type of behavior is happening among households.

In Chap. 9 we sum up our findings and set out the rationale for our proposed CUT policy model for driving forward heating fuel savings in homes in Germany. 'C' is for allowing cost-effective thermal upgrade measures: usually modest, low-gain measures, which nevertheless do pay for themselves within a few years. 'U' is for stimulating user behavior: studies are needed to find out why certain households choose to consume significantly less heating energy than the norm and how they have managed to put effective strategies in place to achieve this—and what sort of campaign and promotional tools would be needed to foster this type of motivation and day-to-day behavior in other households. 'T' is for encouraging—though not subsidizing—top-end thermal retrofitting, such as is currently the legal standard and beyond. As this is not generally 'economically viable' (in the natural meanings of these words), it seems futile for policymakers to make economic viability the main plank of its promotional drive. There may be other good reasons that a large sector of homeowners should be retrofitting to top-end standards: it can make homes more comfortable; provide a buffer against future uncertainties in fuel price; save CO_2 (even if not economically optimally); and modernise a home. In our tentative but conservative calculations we suggest that a well-balanced, three-pronged policy thrust based on CUT streams could reduce heating fuel consumption in German households by 2050 by around 34%, compared to the 9.6% we estimate is likely from the current policy regime.

1.6 Conventions and Variables Used in this Book

German literature uses two main metrics for 'whole-building' thermal performance. The first, 'transmission losses', is a measure of the rate of heat energy loss through the building envelope. It is symbolized by H_T and measured in watts per square meter of the building envelope per degree Kelvin difference between indoor and outdoor temperature (W/m^2K). The performance of individual components of the building envelope (windows, walls, roofs, doors) is given in U-values, which have the same units. These are purely objective measures that are not affected by what the people in the house do. They are simple descriptions of the thermal properties of materials and components.

The second measure of performance is the estimated quantity of heat energy consumed by the building in its year-round running. This is affected by H_T, above, together with the type and efficiency of the boiler; the buildings' air-tightness; its orientation to the sun; the average annual weather conditions; and the way the people in the house heat and ventilate it. The German Institute of Standards *(Deutsches Institut für Normung—DIN)* sets out standard assumptions for occupants' behavior, so that the thermal performance of different buildings may be compared on a quasi-objective basis. This *quantity* of energy consumed is given in relation to the floor area of the building. The units are kilowatt-hours consumed per square meter of floor area per year (kWh/m^2a). In keeping with German practice we use the symbol 'a' for 'year' in this expression (from the Latin *annus*).

Since the EnEV came into force in 2002, this latter measure has to be given in terms of *primary* energy, i.e. it has to include the energy consumed to get the fuel to the house. It is usually given the symbol Q_P, a practice we adopt in this book. However, as this is the *theoretical, calculated* consumption (based on the DIN assumptions) rather than the *actual, measured* consumption, we use another term, Q_M, to denote the actual, measured primary energy consumption.

Further, there are two main measures of floor area in Germany. 'Living area' *(Wohnfläche)* refers to the area inside the front door that is used in normal, daily living. 'Usable area' *(Nutzfläche)* is larger: it includes apartment hallways, attics, and basements, and in detached and semi-detached houses it includes basements and undeveloped (i.e. non-liveable) lofts. The quantity Q_P is always given in terms of 'usable area' in legal documents such as the EnEV and in most empirical studies on building performance, and we keep to that convention in this book. It should be noted, therefore, that a house with a Q_P of, say, 70 kWh/m^{2}a (the current maximum permitted for new builds) actually consumes significantly more energy than that per square meter of living area: 95 kWh/m^{2}a if we use the legal formula, for 'small' buildings, of *Nutzfläche* $= 1.35 \times$ *Wohnfläche*.

Finally, Q_P is always a combined measure of space and water heating consumption. In keeping with this convention we use the term 'heating' to mean space and water heating throughout this book, except where otherwise indicated.

References

Beerepoot M, Sunikka M (2005) The contribution of the EC energy certificate in improving sustainability of the housing stock. Environ Plann B 32(1):21–31

Faber J, Schroten A (2012) Behavioural climate change mitigation options and their appropriate inclusion in quantitative longer term policy scenarios—main report. CE Delft, Delft. http://ec.europa.eu/clima/policies/roadmap/docs/main_report_en.pdf. Accessed 19 Dec 2012

Galvin R (2010) Thermal upgrades of existing homes in Germany: The building code, subsidies, and economic efficiency. Energy Build 42(6):834–844

Galvin R (2011) Discourse and materiality in environmental policy: the case of German federal policy on thermal renovation of existing homes. PhD thesis, University of East Anglia. http://justsolutions.eu/Resources/PhDGalvinFinal.pdf. Accessed 19 Dec 2012

Galvin R (2012) German federal policy on thermal renovation of existing homes: a policy evaluation. Sustain Cities Soc 4:58–66

Galvin R (2013) Impediments to energy-efficient ventilation of German dwellings: a case study in Aachen. Energy Build 56:32–40. doi:10.1016/j.enbuild.2012.10.020

Hargreaves T, Nye M, Burgess J (2010) Making energy visible: a qualitative field study of how householders interact with feedback from smart energy monitors. Energy Policy 38:6111–6119

Meijer F, Itard L, Sunikka-Blank M (2009) Comparing European residential building stocks: performance, renovation and policy opportunities. Build Res Inf 37(5):533–551

Shove E, Chappels H, Lutzenhiser L, Hacket B (2008) Comfort in a lower carbon society. Build Res Inf 36(4):307–311

Sunikka M (2001) Policies and regulations for sustainable building, a comparative analysis of five European countries. Delft University Press, Delft

Sunikka M (2003) Fiscal instruments in sustainable housing policies in the EU and the accession countries. Eur Environ 13(4):227–239

Sunikka M, Boon C (2002) Environmental efforts in the Netherlands social housing sector. Delft University Press, Delft

Sunikka M, Boon C (2003) Environmental policies and efforts in social housing: the Netherlands. Build Res Inf 31(1):1–12

Sunikka M, Boon C (2004) Introduction to sustainable urban renewal: CO_2 reduction and the use of performance agreements: experience from the Netherlands. Delft University Press, Delft

Sunikka M (2006a) Policies for improving energy efficiency of the European housing stock. IOS Press, Amsterdam

Sunikka M (2006b) Energy efficiency and lower carbon technologies in urban renewal. Build Res Inf 34(6):521–533

Sunikka-Blank M, Chen J, Dantsiou D, Britnell J (2012) Improving energy efficiency of social housing areas—a case study of a retrofit achieving an 'A' energy performance rating in the UK. Eur Plann Stud 20(1):133–147

Chapter 2
Development of German Retrofit Policy

Abstract Since 2002, prescribed thermal standards have been mandatory for homes being renovated in Germany. This policy has become entwined with Germany's climate policy of 80% reductions in GHG emissions by 2050, and is the main tool used by the Federal government to drive forward home heating energy efficiency improvements. A second policy instrument is Federal subsidies, given to thermal upgrades that are designed to do better than the standards demanded in the regulations. Thirdly, the Government actively promotes thermal retrofits by promoting the view that the standard of thermal upgrade demanded by regulations always pays back, through heating fuel savings, and therefore costs nothing in the long run. Further policy instruments are energy performance certificates and free on-site advice to would-be home renovators, though there is no building inspection to check whether standards have been reached. The annual rate of thermal retrofit is a disappointing 0.8–1.0% of the housing stock, while average fuel savings per retrofit project are around 25%, so progress falls well below what is needed to reach the 80% goal.

Keywords Thermal retrofit · Housing stock · German policy · Building regulations · Retrofit subsidies

2.1 Introduction

Unlike many other countries, which rely on voluntary measures and market mechanisms to bring about energy and carbon savings in the residential sector (see Chap. 3), the German Federal government uses a mandatory approach as its main policy instrument. The main Federal policy instrument is the *Energieeinsparverordnung* (Energy Saving Regulations: commonly abbreviated 'EnEV'). The EnEV, introduced in 2002 and most recently upgraded in 2009, prescribes the thermal standards to which buildings have to conform. It covers both residential and non-residential buildings, and although the rules for both are comparable, they

R. Galvin and M. Sunikka-Blank, *A Critical Appraisal of Germany's Thermal Retrofit Policy*, Green Energy and Technology, DOI: 10.1007/978-1-4471-5367-2_2, © Springer-Verlag London 2013

are not identical. In this book, we focus specifically on residential buildings. The EnEV covers both new-builds and the retrofit of existing buildings, though again, the rules for each of these are different.

The stated aim of the EnEV is to help mitigate climate change. This is affirmed, for example, in the preamble to EnEV 2009 as presented to the German Upper House of Parliament (the *Bundesrat*):

> The aim of the *Energieeinsparverordnung* is to reduce the energy consumption in the built environment to a sustainable level. As a consequence of the reduction of energy use, fossil fuels will be saved and the emission of climate-damaging greenhouse gases will be significantly reduced (BR 2008: A; authors' translation).

Other Federal documents and declarations quantify this as an 80% reduction in GHG emissions from buildings by 2050 compared to 1990 levels (UBA 2007: 2; BMU 2007: 4–6; Tiefensee 2006). Since GHG emissions from home heating in Germany are almost exactly proportional to energy consumption, this implies the need for an 80% reduction in home heating energy consumption. From the outset, then, we see that two different strands of German government policy have been merged together in the thermal building regulations: its climate goal of reducing GHG emissions from all sources by 80% by 2050 compared to 1990 levels, and the goal of improving the thermal quality of the housing stock.

Of course, the thermal quality of the German housing stock has improved since 1990. But the number of households has increased by 16%, so that the net reduction in heating energy consumption since 1990 is only 4% (BMVBS 2011). As we show in Chap. 7, given the expected continued rise in the number of households, Germany will still need an average reduction of 80% in all the housing stock existing in 2012, to reach the 80% reduction goal.

This chapter describes Germany's thermal retrofit policy, focussing on the key policy instruments, explores how well the EnEV is achieving the Federal government's aims for it and, if it is failing to meet these aims, asks why this might be.

Section 2.2 outlines the development of Federal regulations for thermal retrofits of new builds and existing homes. Section 2.3 describes the Federal subsidy system for thermal retrofit projects. Section 2.4 explains how the government attempts to use the notion of 'economic viability' to motivate homeowners to retrofit their homes, while Sect. 2.5 looks at other policy levers. Section 2.6 gives a brief account of progress so far in reaching national fuel saving goals through thermal retrofits, and Sect. 2.7 concludes.

2.2 The German Thermal Building Regulations

Federal policymakers explain that there are three broad types of policy instruments they employ to induce homeowners to thermally retrofit their buildings: demand; incentivise; and inform (*fordern; fördern; informieren*) (Galvin 2012). The first of these comes in the form of regulations. Germany first included thermal standards in its building regulations in 1977, in the wake of the oil crisis. The standards have been tightened four times since then, and the formulas for working out maximum permissible heat energy consumption have developed through several phases. In broad terms, a building constructed today may consume approximately 25% of the heating energy which a comparable building constructed in 1977 could consume. However, a building undergoing a comprehensive thermal retrofit today may consume about 40% of what a 1977 new build could consume. Nevertheless, a building undergoing a *partial* thermal retrofit today, such as a new roof or major repairs to a wall, must achieve new build thermal standards for the part undergoing retrofit.

The process of development of these regulations illustrates both the learning that German policymakers have undergone, and the dilemmas inherent in trying to set universal thermal standards for buildings. What follows is a simplified account of what has in fact been an extremely complicated process.

The thermal standards set in 1977 applied to new builds and reconstructions only, and were set down in the *Wärmeschutzverordnung*, or 'heat retention regulations', abbreviated as WSVO. These specified the maximum heat transmission loss through the building envelope as a whole. This is called the 'U-value', expressed in Watts per square metre of the building envelope per degree Kelvin difference between indoor and outdoor temperature (W/m^2K). The permissible U-values ranged from 0.91 W/m^2K for buildings with a large surface area relative to their volume (e.g. small or oblong buildings), to 1.40 W/m^2K for buildings with a smaller surface area relative to their volume (e.g. large or more cubic-shaped buildings) (WSVO 1977). Since buildings with large surface area and small volume are difficult to heat efficiently, they were required to have better insulation, i.e. a lower rate of heat transfer through the walls.

In 1982, these values were tightened to a range of 0.6–1.2 W/m^2K (WSVO 1982). Table 2.1 shows the development of thermal requirements over the last 35 years.

In 1995, the standards were tightened further, and a more complex formula was introduced (WSVO 1995). First, U-values were no longer specified for the building

Table 2.1 Average legal maximum primary energy consumption for space and water heating in new buildings; maximum heat consumption and transmission losses for new builds and retrofits

Ordinance	Date in force	Average Q_P (kWh/m²a) for new builds	Average H_T (W/m²K) for new builds	Wall insulation thickness for new builds (cm)	Wall insulation thickness for partial retrofits (cm)	Wall insulation thickness for full retrofits (cm)
WSVO	1977	265	0.96	3	N/A	N/A
	1982	220	0.78	5	N/A	N/A
	1995	150	0.68	8	N/A	N/A
EnEV	2002	100	0.60	12	12	8
	2009	70	0.48	16	16	12

Sources Hegner (2009), plus author calculations based on WSVO and EnEV texts, assuming polystyrene wall insulation of $\lambda = 0.035$ W/mK

envelope as a whole, but for its constituent parts. Second, a building had to comply with a specified maximum permissible 'heating energy consumption' for its space heating and hot water. Called Q_P in the EnEV and now in the building industry in Germany, this was set in relation to the 'useful area' of the building, and expressed in kilowatt-hours per square metre per year (kWh/m^2a). The units kWh/m^2a have become a widely used measure of building performance internationally, and are used extensively in this book. Unlike transmission losses, heating energy consumption is determined by far more factors than merely the quality of the building envelope. These include the efficiency of the heating system, the solar gain through the windows, the way the building is ventilated and, most important, the way the occupants behave. Occupant behaviour is extremely varied, as we shall see in Chap. 5, so it was necessary to assume a standardised set of behaviours. This consists of keeping the entire building at a constant 19 °C all winter, and ventilating the building at the constant rate of one complete change of air every one-and-a-half-hours (0.7 V/h).

The average Q_P permitted for new builds in the WSVO of 1995 was around 150 kWh/m^2a. This compared to an equivalent of 265 kWh/m^2a in 1977 and 220 kWh/m^2a in 1982 (Hegner 2009). Of course, the figure for Q_P calculated by this method was an entirely theoretical number, which was not intended to reflect the way buildings actually perform with real people in them. However, it was such a convenient measure of the overall thermal performance of a building that it rapidly came to be used in Germany and many other countries as a standard. In Germany, it is usually the figure on a building's energy performance certificate (EPC), and has been treated, over the last decade, as an objective measure. As we shall see in Chaps. 5, 6 and 7, this can lead to miscalculations of the energy actually saved when old buildings are thermally retrofitted.

In 2002 the Social Democrat-Green coalition government replaced the WSVO with the *Energieeinsparverordnung,* or 'energy saving regulations', abbreviated as EnEV (EnEV 2002). Again the calculation methods were changed. The Q_P requirement was retained, but the older system of one overall U-value for the entire building envelope, now called H_T, was reinstated alongside this, and a number of rather abstruse complications in the WSVO of 1995 were removed. The average for Q_P was now 100 kWh/m^2a, and that for H_T was 0.60 W/m^2K. The smallest new buildings, with the least thermally efficient geometry, could consume up to 150 kWh/m^2a, but had to have an H_T value no higher than 0.44 W/m^2K. The largest, with the most thermally efficient geometry, could consume only 80 kWh/m^2a but could have the less stringent H_T value of 1.05 W/m^2K. In this way, the EnEV offered a neat compromise: smaller buildings had to have thicker insulation to offset their thermally inefficient geometry, but not to the extent of requiring them to match the energy consumption performance, per square metre of living area, of larger buildings.

A significant feature of 'EnEV 2002', as it came to be called, was that for the first time compulsory thermal standards for retrofits were introduced. Whenever 20% or more of any feature of a residential building (such as a wall or roof) was being repaired or renewed, that entire feature had to be thermally renovated to the same standard as a new build. The U-values achieved for, say, a renovated wall or roof, had to be at least as good as those for a new build. However, if an entire residential building was being comprehensively renovated, the overall thermal standard (Q_P) was permitted to be 40% less stringent than the new build standard. This provision recognised that, while a new build can be designed from scratch with thermal efficiency in mind, existing buildings often have practical limits on the thickness of insulation that can be applied, and cannot be rotated to face the sun.

A further feature of EnEV 2002 was its stated aim as an instrument for climate change mitigation. Its preamble declared that it 'provides a significant element of the government's climate protection programme' and could be expected to bring about 'a 25% reduction in CO_2 emissions from the built environment by 2005 compared to 1990 levels' (Begründung der EnEV 2009, paragraph 1; authors' translation).

Finally, EnEV 2002 stated that the thermal standards it demanded were 'economically viable' *(wirtschaftlich)*, in that the costs they incurred would always pay back, through fuel savings, over the technical lifetime of the thermal retention measures. Prior to the setting of Q_P and H_T values in EnEV 2002, the Housing Ministry (*Bundesministerium für Verkehr, Bau und Stadtentwicklung*—BMVBS) had commissioned an expert report to establish how strict these values could be, without the thermal features of a new build or retrofit costing more than the money its owner would get back through fuel savings. Some of the more interesting and controversial features of this report (Feist 1998), and the thinking that lay behind it, will be discussed in detail in Chaps. 6 and 7.

Nevertheless, a clause was inserted in EnEV 2002 to allow for exceptions to the rules in cases of 'unreasonable expense' or 'intolerable difficulty' (EnEV 2002: paragraph 25). In such cases, a homeowner may apply to the local municipal authorities, who have the power to grant such an exception. The onus is on the homeowner to prove that reaching the EnEV thermal standard would not be economically viable.

A new version of the EnEV came into force in 2009 (EnEV 2009). This tightened the thermal requirements for new builds and retrofits by a further 30%, so that the average Q_P for new builds was now 70 kWh/m²a and for comprehensive retrofits 100 kWh/m²a. Secondly, retrofits now had to conform to EnEV standards if only 10% of a feature of the building was being renovated. Thirdly, every new build must include some form of on-site renewable energy generation.

Once again, the methodology for working out the thermal features of a building was changed, now based on a complex interplay of Q_P, H_T and the thermal features of individual building elements. In brief, the U-values for individual components (walls, windows etc.), together with the predicted energy gain from renewables and sun-facing windows, must produce a Q_P value no higher than that which would be produced by a building of exactly the same design but with U-values, for specific components, given in a table in the regulations (see Box 2.1). This allows more flexibility of design. For example, it allows more windows in a building. Windows have inherently high transmission losses, and there is a temptation to seek to achieve a low Q_P value by having more wall area and less window area. However, to prevent designers producing glasshouse dwellings, a further restriction applies: there is a maximum overall H_T value to which the building must conform. This value varies depending on the number of external walls a building has: it is lowest (most stringent, at $H_T = 0.4$) for detached houses, and highest (least stringent, at $H_T = 0.65$) for apartment blocks sandwiched between other buildings.

The EnEV was due to be further updated in September 2012. Until shortly before this time, the Government's stated intention was to tighten the standards for both new builds and retrofits by a further 30% in 'EnEV 2009'. However, our discussions with Federal policymakers in January 2012 indicated that there were now serious reservations as to whether it would be wise to do so for retrofits. There were clear signals from the building industry that EnEV 2009 had already pushed retrofit costs beyond what could reasonably be regarded as economically viable. Our previous research had suggested to Federal policymakers in 2011 that this was a serious issue, based on an extensive review of existing case studies together with new case studies of our own (Galvin 2010, 2011, 2012). Hence we were not surprised when it was announced in summer 2012 that the standards for thermal retrofits would not be tightened in 2012, nor for the foreseeable future. Interestingly, it was also decided to delay tightening new build standards until 2014, and by 12.5% rather than the 30% originally envisaged (BMVBS 2012).

Box 2.1 Designing a Building with EnEV 2009

The designer makes a computer model of the intended building using a table of maximum permissible U-values and minimum permissible renewable energy. The computer generates a maximum permissible Q_P value for this building. The designer then rebuilds the model using the renewable energy capacity and U-values of the *actual* components she wants to use, and checks what Q_P value this generates. If this is higher than the original Q_P, she has to change her building components until her Q_P comes down the maximum permissible level. Finally, she checks that the building conforms to the H_T value for its type (detached, semi-detached, etc.). The computer program is based on a calculation methodology published by the German Institute of Standards (*Deutsches Institut für Normung*—DIN), referred to in the EnEV text. The same procedure is followed for comprehensive retrofits, but the Q_P value may be 40% higher.

2.3 Federal Subsidies

The second instrument Federal policymakers use to induce homeowners to thermally retrofit their buildings is financial incentives (see Chap. 6). Federal subsidies are given as loans or grants, to both home refurbishment and new build projects, via the German Development Bank (KfW—*Kreditanstalt für Wiederaufbau*). The cluster of incentives within this programme comes under the broad heading of the CO_2-*Gebäudesanierungsprogramm* (CO_2-building refurbishment programme).

The KfW programme started in 2001 and, with respect to existing homes, initially supplied subsidised loans only, and only where a building was being comprehensively refurbished. Later this was broadened to include partial refurbishments, and as from 2009 homeowners were given the choice of a subsidised loan or an outright grant, with both forms of subsidy equivalent in terms of their net present value.

To qualify for subsidies, a comprehensive refurbishment project has to be designed to achieve energy consumption (Q_P) that betters the EnEV *refurbishment* standard for that project by 20% or more, i.e. it can consume 115% of the EnEV 2009 *new build* standard (which is 40% more stringent than the refurbishment standard)[1]. Hence, the minimum qualifying standard is called the 'KfW-115 house'. This demands an average of 80 kWh/m^2a, though as we saw above, the actual EnEV standard for each project depends on the building's geometry and volume. The level of subsidy increases according to the percentage by which the project beats the EnEV standard, up to a maximum subsidy for projects that beat it by 60% or more—an average of 40 kWh/m^2a. There is also an additional requirement, that projects must beat the transmission loss standard (H_T) for retrofits, by a corresponding percentage. Table 2.2 shows the standards demanded for various levels of subsidy, as published by the KfW. The highest standard is the 'KfW-55 house', which consumes 55% of the new-build standard for a building of the same geometry, volume and design.

We may ask why the subsidies are only given to projects that exceed the legal standards. Policymakers maintain this is because thermal retrofit projects which merely reach the legal standard will pay for themselves, over the technical lifetime of the refurbishment measures, through reduced expenses for heating fuel. As mentioned above, the government maintains that the standards demanded in the EnEV are 'economically viable': they have been set at such levels as to fit neatly with current costs of thermal refurbishment. Hence, it is argued, it does not actually cost a homeowner anything to refurbish to EnEV standard, as the thermal improvement will pay for itself. However, it is argued, it does cost to go beyond this standard, because at these levels of thermal refurbishment the costs escalate and the return per euro diminishes considerably (cf. Galvin 2010; Jakob 2006). Hence, the subsidies are given only for thermal improvements above and beyond

---

[1] The arithmetic is non-intuitive: 80% × 1.4 ≈ 115%.

Table 2.2 Levels of subsidies and thermal standards required for KfW subsidies for retrofit projects

German name of standard	Q_P as percentage of new build standard	H_T as percentage of new build standard	Average Q_P (kWh/m²a)	Improvement on average EnEV retrofit standard	Subsidy as % of thermal costs	Maximum subsidy
KfW-Effizienzhaus 55	55%	70%	40	60%	20%	€15,000
KfW-Effizienzhaus 70	70%	85%	50	50%	17.5%	€13,125
KfW-Effizienzhaus 85	85%	100%	60	40%	15%	€11,250
KfW-Effizienzhaus 100	100%	115%	70	30%	12.5	€9,375
KfW-Effizienzhaus 115	115%	130%	80	20%	10%	€7,500

Source KfW (2012)

the EnEV requirements, while the homeowner pays for the improvements up to the EnEV level.

An important consequence of this is that the KfW subsidies are only financing the most economically inefficient extremes of thermal refurbishment. Federal politicians and bureaucrats argue, however, that the offer of KfW subsidies induces homeowners to embark on refurbishment projects by offering them the incentive of something for nothing (see policymaker interviews in Galvin 2011). Hence, they argue, the few billion euros in the KfW annual budget 'trigger' (*auslösen*) the spending of tens of billions of private capital on refurbishment that incorporates energy-efficiency improvements. More will be said about this in evaluating the cost effectiveness of thermal retrofit in Germany, which will be discussed in Chap. 6.

2.4 The Economic Viability of Thermal Retrofits

As we noted above, the EnEV may only demand thermal standards that are 'economically viable'. This was originally a provision to protect homeowners from being forced to pay inordinate sums to incorporate unreasonably stringent thermal upgrade measures in their retrofits or new builds. However, it has taken on a new function, in policy discourse, as a motivational and promotional tool to persuade homeowners to undertake thermal retrofits who might otherwise not have done so: it costs nothing in the long-run, so they should do it. It is important to understand the legal meaning of the term 'economically viable', and how this affects its usefulness as a motivational tool.

The EnEV is a set of regulations. Although changes in the EnEV have to be approved by both Federal houses of parliament, they also have to conform to a law which sets the EnEV's boundaries and limits, namely the *Energieeinsparungsgesetz* (energy saving law—'EnEG'; see EnEG 2009). This stipulates (paragraph 5) that energy saving measures for buildings may only demand thermal upgrade measures that are *'wirtschaftlich'* (economically viable). In the German policy context, this is interpreted to mean three things.

First, the savings expected to be won as a result of the thermal upgrade measures have to be great enough to pay back the cost of those measures, over their technical lifetime. This is usually taken to be 25 years, but it can vary between components (walls, roofs, boilers, windows etc.) and types of building and insulation material. For example, if the thermal features of a retrofit project cost €15,000 and this is expected to result in fuel savings worth €650 per year over 25 years, the project would be deemed economically viable, since €650 × 25 = €16,250. Of course, savings will vary depending on vacillations in the fuel price, while costs will vary depending on the interest rate being paid if the homeowner took out a loan to finance the project. Other factors also come into the calculation, and in Chaps. 6 and 8 we look more closely at the mathematical models used to determine economic viability.

Second, in the calculation of economic viability, only the costs of the thermal upgrade measures are taken into account (the *energetische Mehrkosten*—'thermal improvement costs'), and not the costs of associated structural changes or maintenance measures (the *sowieso-Kosten*—'anyway costs'). So, for example, in calculating whether an external wall insulation job will be economically viable, the regulators only have to consider the costs of the insulation material and the labour of fixing it to the wall (the 'thermal improvement cost'). They do not have to take into account the costs of scaffolding, stripping the old render, applying a new render and painting the finished wall (the 'anyway costs'). It is argued that the latter would have to be paid anyway, in the normal course of building maintenance. The legal definition of economic viability assumes that anyone doing a thermal upgrade would do so during major maintenance or retrofit, the costs of which would have to be paid in the normal course of building maintenance. Before each tightening of the EnEV standards for retrofits, in 2002 and 2009, the government commissioned expert reports (i.e. Feist 1998; Kah et al. 2008) to confirm whether the intended standards were economically viable according to these criteria. The author of the report for EnEV 2002, who was also the driving force behind the report for EnEV 2009, continues to affirm this principle (Feist 2009), and it is widely accepted as right and proper in the policymaking community (Galvin 2011, 2012).

The advantage of this approach is that it integrates thermal retrofit into the regular and incidental maintenance of residential buildings. Although homeowners are required, by regulation, to include thermal upgrading in such maintenance projects, they are also assured that, in the long run, this will pay back.

There is a difficulty here, of course, for homeowners whose buildings are not due for major maintenance, but who wish to upgrade them thermally. They have to conform to EnEV thermal standards, but all their costs, from their point of view, are for the thermal upgrade.

A further difficulty is that not all homeowners are likely to regard their buildings as due for major maintenance within the next 38 years, i.e. up until 2050. If the regulations effectively force homeowners to wait until major maintenance is due, before they can afford to thermally upgrade to the required legal standard, many will wait for a very long time. The policymakers we have interviewed counter this with the remark that all pre-1977 German homes are in need of major maintenance, and those built between 1977 and 1994 will be before 2050. Our interviews with homeowners indicate, however, that for them the question as to when a home is due for major maintenance is a subjective one, and has as much to do with household budget as to the appearance and condition of the building.

Third, the calculation of economic viability is based on a building's *theoretical* thermal performance, not its *actual, measured* performance before and after retrofit. There can be a great difference between these two. As we will discuss in Chap. 5, the actual, measured heating fuel consumption in German dwellings is, on average, about 30% lower than the theoretical consumption (Sunikka-Blank and Galvin 2012), and the gap is even wider for older, thermally inefficient dwellings. The main reason for this seems to be that households in thermally leaky buildings often try to avoid heavy fuel bills by adapting to cooler indoor temperatures, rationalising the use of rooms, or turning radiators on and off strategically to suit their room use. An important effect of this gap is that a retrofit job that looks economically viable on paper might not be in the real world: one cannot save energy one is not already consuming (see Chap. 6).

Despite these difficulties, the Federal government uses economic viability as a key idea in promoting thermal retrofit. Typical is a full-page advertisement in a promotional magazine issued free with major German weekend newspapers on 15–16 September 2012 (Bundesregierung 2012, p. 7). The headline declared that thermal retrofit of buildings *'lohnt sich'*, a word-play meaning both 'pays for itself' and 'is worthwhile'. This reflects the Government's ongoing promotion of thermal retrofits on the grounds that these are economically viable. Further examples may be seen on the websites of the Housing Ministry (BMVBS— www.bmvbs.de) and the German Energy Agency (DENA—www.dena.de).

2.5 Further Policy Levers: Energy Advice, Energy Certificates and Demonstration Projects

An important instrument for promoting thermal retrofit of existing homes in Germany is the role of local, on-site energy advisors. Homeowners considering a thermal retrofit can arrange for a visit from a qualified energy advisor, who will examine the building, discuss their needs, and suggest various alternatives for thermal retrofit and their likely costs and benefits.

The tasks of an energy advisor are exacting, as she or he must assess the thermal quality and retrofit potential of a dwelling in a short visit, and also have the personal skills to engage with a household in exploring their needs and aspirations for their home. There are now national and state training schemes in place.

We have not found any studies of the effects of these visits, but our own interviews with homeowners suggest they generally help people understand the thermal and economic issues. Whether they induce people to undertake major thermal retrofit projects is another matter. Some retired interviewees, for example, said the visit of a local, municipally funded energy advisor helped convince them that extensive thermal retrofit would not be economically viable for them, as they were unlikely to live out the 25 years required for payback, at least in their present home.

The Ministry for the Economy and Technology (*Bundesministerium für Wirtschaft und Technologie*—BMWi) now funds energy advisors' visits, but only under the condition that they advise the homeowner to renovate in such a way that the dwelling will come up to the full EnEV standard or better (BMWi 2012). Hence, Federally paid energy advisors are under some duress to give advice that suits government policy, rather than advice that their own skills and experience might indicate is best for the actual constellation of building and household.

A further policy instrument is the Energy Performance Certificate (EPC), a requirement stemming from EU rules. The Energy Performance of Buildings Directive requires that 'when buildings are constructed, sold or rented out, an energy performance certificate is made available to the owner or by the owner to the prospective buyer or tenant ...' (EC 2003, p. 68, Article 7). In Germany, all new buildings must have an EPC. An existing dwelling must have one if it is being offered for sale or rent, and also both before and after a thermal refit. Landlords and vendors do not have to display the EPC, but must produce it when asked by prospective tenants or purchasers. EPCs display the energy consumption of the dwelling, in kWh/m²a, and also the options for a thermal upgrade. Generally the consumption displayed is the theoretical, calculated value (in German the *Energiebedarfskennwert*). However, for certain categories of buildings (see Chap. 5, Sect. 5.1, if a household has three continuous years of energy bills their dwelling's EPC may instead display the actual, measured consumption (in German the *Energieverbrauchskennwert*). As noted above, the average gap between these two measures is 30%.

The EPC is intended to motivate homeowners to renovate, as it is assumed that a low energy consumption figure increases the market value of a property. Its effectiveness in this regard has been investigated in recent studies in Germany (BMVBS 2011; Amecke 2011) and the UK (Adjei et al. 2011; Lainé 2011). Both UK studies found that homebuyers rarely used the EPC in assessing the worth of a home or negotiating the sale price. The smaller of the two German studies (n = 151), BMVBS (2011), found that 42% of respondents felt the certificate provides useful information about a building, though 30% said they would prefer it not to be there, as it weakened their negotiating position. This is an indication that, for these cases at least, a good energy rating can influence the value of the property.

The larger German study (n = 1,239), Amecke (2011), found that 35% of purchasers viewed the EPC for a property they were closely considering, and the certificate was actively promoted to 24% by the vendor or agent. Most found the certificates understandable, but only 44% regarded them as trustworthy. Further, the metric on the certificate (kWh/m²a) was meaningless to most purchasers. Although most rated the cost of energy as a factor in a decision to purchase, most found energy bills, site visits and professional advice far more useful than the EPC. Finally, as a factor influencing their purchasing decision, energy efficiency was 9th in order of importance behind other factors mentioned in the survey.

Hence there is some evidence that an EPC can influence negotiations between vendor and purchaser, but not enough to make it a strong motivating factor for undertaking thermal retrofit. Other factors, such as location, layout, décor and the quality of indoor fittings appear to play a much larger role in home purchasing decisions.

Finally, a policy instrument notable by its absence in Germany is inspection of homes that have been newly built or renovated. While Germany has strict regulations for both thermal and structural features of buildings, there is no provision for inspection of projects to check their conformity to the law. This has interesting effects. First, there is no systematic feedback of the effectiveness of thermal retrofits that are designed to reach or do better than the EnEV requirements. Instead, the Government relies on random surveys, commissioned by the Housing ministry, to assess the energy savings achieved in projects that receive KfW subsidies (e.g. Clausnitzer et al. 2010), and on expert reports to assess the progress of thermal retrofits overall (e.g. Friedrich et al. 2007).

Second, the lack of inspection creates an incentive for homeowners to renovate below the thermal standard required by the EnEV. We have seen many examples of this in Germany, some of which are outlined, with locations changed to preserve anonymity, in Galvin (2011). From the perspective of the potential of the building to be thermally upgraded, some of these seem quite sensible, such as a Brandenburg homeowner who put 6 cm of roof insulation under new roof tiles rather than the regulation 22 cm, to avoid having to restructure her roof. Others seem like a wasted opportunity, such as a homeowner in northern Bavaria who put new render on the walls of his house—complete with the expense of scaffolding—without adding any insulation at all.

Third, and despite many wasted opportunities, we see some evidence that the lack of inspection enables some homeowners to do at least a degree of thermal upgrade, where the EnEV requirements are too strict, in their case, to be affordable. There is an ongoing discussion among policymakers as to whether 'broader is better than deeper': will more energy be saved by allowing modest, inexpensive thermal up grades, or by restricting thermal upgrades only to strict standards that are inevitably expensive? We will return to this point in Chaps. 6 and 9.

2.6 Achievements and the 80% Goal

Throughout Germany one sees examples of residential buildings renovated to high thermal standards. The German Energy Agency (*Deutsche Energie-Agentur—* DENA) keeps a database of demonstration projects to show that such standards are possible to achieve in a diverse range of existing buildings of all ages (www.dena.de). Two types of studies track the effectiveness of Germany's thermal retrofit policy: those tracking the annual rate of thermal retrofit, and those investigating the depth of fuel savings achieved. In Chap. 6 we will consider these in detail. At this point we merely note the main findings.

Studies of the annual rate of thermal retrofit put this at between 0.8 and 1.0% of the housing stock (Diefenbach et al. 2010; Friedrich et al. 2007; Tschimpke 2011; Weiss et al. 2012). Meanwhile, studies suggest that the average fuel consumption reductions being achieved in the more ambitious retrofit projects—those receiving KfW subsidies—are around 33% per project (Clausnitzer et al. 2009, 2010; Diefenbach et al. 2011). This is a theoretical figure, based on a calculated assessment of the thermal quality of buildings pre- and post-retrofit. When the actual, measured pre- and post-retrofit consumption is considered, the fuel savings fall to around 25% (Michelsen and Müller-Michelsen 2010; Schröder et al. 2011; Walberg et al. 2011; and cf. DENA 2012).

It would seem, then, that thermal retrofits are bringing an annual reduction in home heating energy consumption of around 0.25% (0.1 × 25 = 0.25). Continuing at this rate to 2050 would lead to a 9.5% reduction, considerably less than the 80% goal. Our ongoing discussions with Federal policymakers indicate they have become increasingly aware of the slow progress of fuel consumption reductions through thermal retrofits. They are particularly concerned about the very slow rate of thermal retrofit among buildings with 1–6 dwellings, which make up the vast majority of German homes and are mostly privately owned.

2.7 Conclusions and Implications

German Federal regulations demand that homes being comprehensively renovated reach an average thermal standard of 100 kWh/m^{2}a, which is 40% of what a new building of the same design in 1977 could consume. The individual components of homes being only partially renovated must reach a more stringent standard, equivalent to 70 kWh/m^{2}a. The government and its expert advisors claim these standards are economically viable, in that their costs pay back in fuel savings over their technical lifetime, usually taken to be 25 years. Feedback from the building industry and a growing number of academics has convinced the government that these standards are at the limit of what today's thermal retrofit technology can achieve economically viably, and the government's plans to tighten the regulations further have been abandoned. Subsidies from the KfW are offered for thermal retrofit projects that go beyond the legally required standards, in an attempt to push fuel savings higher.

Thermal retrofit policy is entwined with Germany's climate goal of achieving 80% reductions in GHG emissions by 2050 compared to 1990 levels. However, the savings actually being achieved through thermal retrofit suggest Germany is heading for reductions of around 9.5% in this sector by 2050.

There are several clear reasons for this wide discrepancy. The economic viability criterion is based on theoretical, rather than actual, fuel savings, and is designed to apply only where thermal upgrades are being inserted into maintenance and repairs that would be funded by cyclical maintenance budgets. Further, the mathematical models used to calculate economic viability are of a character that does not dovetail well with ordinary people's budgeting sense. Many buildings also present technical difficulties for problem-free thermal upgrades because of their geometry or design, increasing costs further. Some homes could be thermally renovated economically viably to a lower standard than the regulations demand, but strict requirements make this illegal. For these reasons, the idea that thermal retrofits to EnEV standards pay for themselves has not caught on among homeowners, and fails to motivate the majority to retrofit their homes.

Nevertheless, Germany has made genuine progress in thermal retrofitting in specific spheres. It now has a robust, widespread infrastructure for undertaking thermal upgrades, at costs which are competitive for a high-wage economy.

Government, state and municipal buildings are being thermally upgraded continuously, and progress among large, consortium-owned apartment buildings is considerable. A portion of smaller residential buildings is being given a new lease of life, even if the actual numbers are small.

Finally, the entwining of thermal retrofit policy with the 80% climate goal seems somewhat artificial, and there is no logical reason why thermal upgrades should lead to 80% reductions in heating fuel consumption. Although a 9.5% projected saving over the next 38 years is a long way from an 80% saving, in itself it would be an achievement. A closer examination of key issues, in later chapters, will bring to light avenues through which this savings rate could be considerably improved.

References

Adjei A, Hamilton L, Roys M (2011) A study of homeowner's energy efficiency improvements and the impact of the energy performance certificate. BRE (Building Research Establishment), Hertfordshire

Amecke H (2011) The effectiveness of energy performance certificates—evidence from Germany. Climate policy initiative, Berlin. http://climatepolicyinitiative.org/wp-content/uploads/2011/12/Effectiveness-of-Energy-Performance-Certificates.pdf. Accessed 5 Oct 2012

Begründung der EnEV (2009) German Federal Government website for the energy saving regulations 2009, Stand: 18. April 2008, Anlage2. http://www.enev-online.de/enev/080518_enev2009_begruendung.pdf. Accessed 26 Oct 2010

BMU (Bundesministerium für Umwelt, Naturschutz und Reaktorsicherheit) (2007) Taking action against global warming: an overview of German climate policy. BMVBS, Berlin

BMVBS (Bundesministerium für Verkehr, Bau und Stadtentwicklung), (2011) Evaluierung ausgestellter Energieausweise für Wohngebäude nach EnEV 2007. BMVBS, Berlin

BMVBS (Bundesministerium für Verkehr, Bau und Stadtentwicklung) (2012) Interview mit Minister Ramsauer: "Wohnen muss bezahlbar bleiben", 4 July 2012. http://www.bmvbs.de/SharedDocs/DE/RedenUndInterviews/2012/BauenUndWohnen/bundesminister-dr-peter-ramsauer-im-interview-mit-der-augsburger-allgemeinen-am-04-07-12.html. Accessed 24 Sept 2012

BMWi (Bundesministerium für Wirtschaft und Technologie) (2012) Energiesparberatung vor Ort: Ein Förderprogramm des Bundesministeriums für Wirtschaft und Technologie. BMWi, Berlin. http://www.bmwi.de/Dateien/Energieportal/PDF/energiesparberatung-vor-ort,property=pdf,bereich=bmwi2012,sprache=de,rwb=true.pdf. Accessed 5 Oct 2012

BR (Bundesregierung) (2008) Verordnung der Bundesregierung: Verordnung zur Änderung der Energieeinsparverordnung (Regulations for changes to the energy saving ordinance), Drucksache: 569/08

Bundesregierung (2012) Energie für Deutschland: saubere, sichere, bezahlbare Energie: Deutschland verändert sich. Rostock: Publikationsversand der Bundesregierung, distributed 14–15 Sept 2012

Clausnitzer KD, Gabriel J, Diefenbach N, Loga T, Wosniok W (2009) Effekte des CO_2-Gebäudesanierungsprogramms 2008. Bremer Energie Institut, Bremen

Clausnitzer KD, Fette M, Gabriel J, Diefenbach N, Loga T, Wosniok W (2010) Effekte der Förderfälle des Jahres 2009 des CO_2-Gebäudesanierungsprogramms und des Programms „Energieeffizient Sanieren". Bremer Energie Institut, Bremen

Diefenbach N, Cischinsky H, Rodenfels M, Clausnitzer KD (2010) Datenbasis gebäudebestand datenerhebung zur energetischen qualität und zu den modernisierungstrends im deutschen wohngebäudebestand. Institut Wohnen und Umwelt/Bremer Energie Institut, Darmstadt. http://www.iwu.de/fileadmin/user_upload/dateien/energie/klima_altbau/Endbericht_Daten basis.pdf. Accessed 21 Jan 2012

Diefenbach N, Loga T, Gabriel J, Fette M (2011) Monitoring der KfW-Programme „Energieef-fizient Sanieren"2010 und „Ökologisch/Energieeffizient Bauen"2006–2010. Bremer Energie Institut, Bremen

DENA (Deutsche Eanergi-Agentur) (2012) Energieeffiziente Gebäude. http://www.dena.de/themen/energieeffiziente-gebaeude.html. Accessed 26 Sept 2012

EC (2003) Council Directive 2002/91/EC of 16 December 2012 on the Energy Performance of Buildings. Offical J Eur Communities 1:65–71

EnEG (Energieeinsparungsgesetz) (2009) Gesetz zur Einsparung von Energie in Gebäuden (Energieeinsparungsgesetz - EnEG), Anfertigungsdatum: 22.07.1976; geändert durch Art. 1 G v. 28.3.2009. http://www.gesetze-im-internet.de/eneg/BJNR018730976.html. Accessed 26 Sept 2012

EnEV (Energieeinsparverordnung) (2002) Verordnung über energiesparenden Wärmeschutz und energiesparende Anlagentechnik bei Gebäuden (Energieeinsparverordnung—EnEV) *) Vom 16. November 2001 (BGBl. I S.3085). http://www.bbsr-energieeinsparung.de/nn_1025172/EnEVPortal/DE/Archiv/EnEV/EnEV2002/2002__node.html?__nnn=true. Accessed 4 Oct 2012

EnEV (Energieeinsparverordnung) (2009) EnEV 2009—Energieeinsparverordnung für Gebäude. http://www.enev-online.org/enev_2009_volltext/index.htm. Accessed 26 Sept 2012

Feist W (1998) Wirtschaftlichkeitsuntersuchung ausgewählter Energiesparmaßnahmen im Gebäudebestand, Fachinformation PHI-1998/3. Passivhaus Institut, Darmstadt

Feist W (2009) Perspectives for the future—passive house technology. In: UNECE conference towards an action plan for energy efficient housing in the UNECE region, Vienna, 23–25 Nov 2009

Friedrich M, Becker D, Grondy A, Laskosky F, Erhorn H, Erhon-Kluttig, Hauser G, Sager C, Weber H (2007) CO$_2$-Gebäudereport 2007, im Auftrag des Bundesministeriums für Verkehr, Bau und Stadtentwicklung (BMVBS). Fraunhofer Institut für Bauphysik, Stuttgart

Galvin R (2010) Thermal upgrades of existing homes in Germany: the building code, subsidies, and economic efficiency. Energy Build 42(6):834–844

Galvin R (2011) Discourse and materiality in environmental policy: the case of German federal policy on thermal retrofit of existing homes. Ph D Thesis, University of East Anglia. http://justsolutions.eu/Resources/PhDGalvinFinal.pdf. Accessed 19 Nov 2012

Galvin R (2012) German Federal policy on thermal retrofit of existing homes: a policy evaluation. Sustain Cities Soc 4:58–66

Hegner HD (2009) Gebäudestandards der Zukunft: die Sicht des Bundesministeriums: Bericht aus Berlin. Public lecture at the Building Physics Department of the Technical University of Munich, 19 Nov 2009

Jakob M (2006) Marginal costs and co-benefits of energy efficiency investments: the case of the Swiss residential Sector. Energy Policy 34:172–187

Kah O, Feist W, Pfluger R, Schnieders J, Kaufmann B, Schulz T, Bastian Z, Vilz (2008) Bewertung energetischer Anforderungen im Lichte steigender Energiepreise für die EnEV und die KfW-Förderung. BBR-Online-Publikation

KfW (Kreditanstalt für Wiederaufbau) (2012) Energieeffizient Bauen und Sanieren. https://energiesparen.kfw.de/html/finanzierungsangebote/energieeffizient-sanieren-430/konditionen/. Accessed 13 Sept 2012

Lainé L (2011) Room for improvement: the impact of EPCs on consumer decision-making. Consumer Focus, London

Michelsen C, Müller-Michelsen S (2010) Energieeffizienz im Altbau: Werden die Sanie-rungspotenziale überschätzt? Ergebnisse auf Grundlage des ista-IWH-Energieeffizienzindex.

Wirtschaft im Wandel, 9, 447–455. http://www.iwh-halle.de/d/publik/wiwa/9-10-5.pdf. Accessed 28 Nov 2011

Schröder F, Altendorf L, Greller M, Boegelein T (2011) Universelle Energiekennzahlen für Deutschland: Teil 4: Spezifischer Heizenergieverbrauch kleiner Wohnhäuser und Verbrauchshochrechnung für den Gesamtwohnungsbestand. Bauphysik 33(4):243–253

Sunikka-Blank M, Galvin R (2012) Introducing the prebound effect: the gap between performance and actual energy consumption. Build Res Inf 40:260–273

Tschimpke O (2011) Auf dem Weg zu einem klimaneutralen Gebäudebestand bis 2050. Naturschutzbund Deutschland (NABU) e. V., Berlin

Tiefensee W (2006) The programme for refurbishment of buildings to reduce CO_2 emissions is a success story of the Federal Government. Press release, Bundesministerium für Verkehr, Bau und Stadtentwicklung (BMVBS), December 2006

UBA (Umweltbundesamt) (2007) Wirkung der Meseberger Beschlüsse vom 23.08.2007 auf die Treibhausgasemission in Deutschland im Jahr 2020 (Implications of the Meseberg Decisions of 23.08.2007 on Greenhouse Gas Emissions in Germany by the year 2020). UBA, Dessau

Walberg D, Holz A, Gniechwitz T, Schulze T (2011) Wohnungsbau in Deutschland—2011 Modernisierung oder Bestandsersatz: Studie zum Zustand und der Zukunftsfähigkeit des deutschen „Kleinen Wohnungsbaus", Arbeitsgemeindschaft für zeitgemäßes Bauen, eV. http://www.bdb-bfh.de/bdb/downloads/ARGE_Kiel_-_Wohnungsbau_in_Deutschland_2011.pdf. Accessed 20 Oct 2011

Weiss J, Dunkelberg E, Vogelpohl T (2012) Improving policy instruments to better tap into homeowner refurbishment potential: lessons learned from a case study in Germany. Energy Policy 44:406–415

WSVO (Wärmeschutzverordnung) (1977) Verordnung über einen energiesparenden Wärmeschutz bei Gebäuden (Wärmeschutzverordnung: Wärmeschutz V) Vom 11. August 1977. http://www.energieberater.de/upload/WSVO%201977.pdf. Accessed 4 Oct 2012

WSVO (Wärmeschutzverordnung) (1982) Verordnung über einen energiesparenden Wärmeschutz bei Gebäuden (Wärmeschutzverordnung—WärmeschutzV) Vom 24. Februar 1982. http://www.bbsr.bund.de/nn_1025190/EnEVPortal/DE/Archiv/WaermeschutzV/WaermeschutzV1982__84/1982__84.html. Accessed 4 Oct 2012

WSVO *(Wärmeschutzverordnung)* (1995) Verordnung über einen energiesparenden Wärmeschutz bei Gebäuden (Wärmeschutzverordnung—WärmeschutzV) *) Vom 16. August 1994.http://www.bbsr.bund.de/nn_1025172/EnEVPortal/DE/Archiv/WaermeschutzV/WaermeschutzV1995/1995__node.html?__nnn=true. Accessed 4 Oct 2012

Chapter 3
German Retrofit Policy in Context

Abstract EU member states are committed to the Energy Performance of Buildings Directive (EPBD) and the Energy Efficiency Directive (EED), including a 20% energy efficiency target by 2020. Most European countries have adopted market-led policies to meet these targets, relying on voluntary take-up by homeowners. Some impose taxes on energy inefficiency, but there is reluctance to do this due to issues of inspection, control, unpredictability, the vulnerability of low-income households, and political sensitivity. Most EU countries offer subsidies for thermal retrofits, though research suggests a weak relationship between subsidies and the diffusion of thermal efficiency technologies. Germany and Finland, for example, rely heavily on technical fixes and deep retrofits, though there are technical and economic difficulties with the depth of thermal improvement demanded. All European policies seem to lack tools to address the energy saving potential of behavioral change, although based on evidence to date occupant behavior can have important implications for the effectiveness of policy instruments.

Keywords Thermal retrofits · Sustainable building · European Union policy · Housing stock · Energy use behavior

3.1 Introduction

Chapter 2 outlined the development of the key elements of Germany's thermal retrofit policy. As Germany is one of the few countries with mandatory thermal requirements for the existing stock and demanding standards for new construction, it is often seen as a global leader in implementing energy efficiency in the building sector (e.g. Sunikka 2006; Pasquier and Saussay 2012; IEA 2008). This chapter sets the German retrofit policy in a wider context, focusing mostly on EU initiatives, and on EU member states with similar economies and climates to Germany (for a wider, global perspective see e.g. de T'Serclaes 2007).

R. Galvin and M. Sunikka-Blank, *A Critical Appraisal of Germany's Thermal Retrofit Policy*, Green Energy and Technology, DOI: 10.1007/978-1-4471-5367-2_3, © Springer-Verlag London 2013

In Sect. 3.2, we consider the two key EU Directives that aim to influence national legislation and policy on thermal retrofitting. In Sect. 3.3, we discuss regulatory approaches to raising thermal standards, compared with market-led, positive, and negative financial incentives in Sect. 3.4. Section 3.5 focuses on the advantages and disadvantages of deep versus incremental retrofits. Section 3.6 discusses household behavior influences on heating energy consumption, including a case study in Cambridge, United Kingdom. Conclusions are drawn in Sect. 3.7.

3.2 European Directives

In March 2007, the EU member states agreed on a 20% energy efficiency target by 2020. This aims to improve energy efficiency by 20%, increase the share of energy from renewable sources to 20%, and reduce EU-wide CO_2 emissions by 20% compared to 1990 levels. This '20-20-20' target aims to integrate climate and energy policy, enhance energy security, and ensure European competitiveness and an increase in jobs. The 20% energy efficiency target is mainly addressed in the implementation of the Energy Efficiency Directive (EED), supported by the Energy Performance of Buildings Directive (EPBD).

3.2.1 Energy Efficiency Directive

The EU reached agreement of the Energy Efficiency Directive (EED) in June 2012. This sets four key requirements for member states to improve their energy efficiency, supporting the EU's 20% target, although in the final negotiation stage the legally binding target was set at 17%. First, energy supply companies (ESCOs) have to reduce their energy sales to industrial and household clients by 1.5% or more each year. This involves introducing clients to energy efficiency measures and encouraging their implementation. A similar type of policy has already been place in some countries. In Finland, since January 2010 a law imposes obligations upon energy companies to provide information to customers on their energy use and on suggested energy efficiency measures. The UK was the first country to introduce a 'Supplier Obligation' (SO) on ESCOs, to save energy at the customer end, in 1994. The policy has been cost-effective, for example in comparison to the German KfW program (see Chap. 2). According to Rosenow (2011), the KfW program and the SO have produced comparable energy savings, but while under the SO about 2 billion euros were spent by energy suppliers between 2002 and 2008, in the same period the KfW spent twice as much, at around 4.5 billion euros of public funds. The difference may partly be due to the higher energy-efficiency standards demanded in the German scheme, which increase the marginal cost of each kWh saved.

Second, the EED sets a target of a 3% annual retrofit rate for public buildings. However, during the lengthy negotiations on the EED's implementation this requirement was weakened, so that it only applies to public buildings that are central government-owned and occupied, for example excluding regional and state-owned properties in Germany. Although this means that only around 30% of Germany's public buildings need to comply, a number of states and municipalities already have their own programmes for thermally upgrading their buildings. Munich, Freiburg, and the city-state of Hamburg are leading examples.

Third, the EED requires each EU member state to draw up a roadmap to make the building sector more energy efficient by 2050, including strategic plans for commercial, public and private households. The roadmaps are intended to provide continuity and security for long-term investment. The commitment to these roadmaps is the responsibility of each national state, but as a positive measure the requirement emphasizes the building sector as one key stakeholder of the policy.

Fourth, the EED includes encouraging additional measures such as energy audits and energy management for large firms, and cost-benefit analysis for the deployment of combined heat and power generation (CHP).

The EED entered into force in November 2012. The member states had to present their national programs for its implementation, together with their national indicative targets, by April 2013. In countries like Germany, which is likely to implement the Directive correctly and in detail, it is likely to have direct impact on energy suppliers, while the national EED roadmap will provoke clarification of the contribution of the Energy Saving Regulations (EnEV), and other policy instruments, to the target of a 20% gain in energy efficiency by 2020.

3.2.2 Energy Performance of Buildings Directive

In early 2003, the European Parliament accepted Directive 2002/91/EC on the Energy Performance of Buildings (EC 2003). One of the four key elements described in the Directive was the introduction of energy certificates for the existing building stock. The directive required that, by January 2006, an Energy Performance Certificate (EPC), not more than 10 years old, must be shown to prospective purchasers or tenants when a new or existing building is sold or let, including recommendations for improvements in energy performance. The directive requires energy certificates to be issued for the existing buildings, but leaves it for each member state to decide whether certain minimum energy criteria should be met, and whether to combine the energy certificate with economic policy instruments (Beerepoot and Sunikka 2005). The UK Government, for example, has announced that it aims to set a requirement that from April 2018 it will be illegal for landlords to rent out homes or business premises with an energy efficiency rating less than 'E', as verified in the EPC. According to the UK Government, at least 682,000 properties are currently below this standard (House of Commons 2011).

The effectiveness of these EU initiatives in stimulating thermal retrofits has been criticized. Aware of these criticisms, the European Commission proposed a recast of the 2002 EPBD as part of its Second Strategic Energy Review in

November 2008. According to the Commission's own estimates, the recast of the EPBD is expected to bring EU energy consumption down by 5–6% and reduce CO_2 emissions by 5% by 2020, all contributing to the 20% energy efficiency target described above. Further, the 2002 EBPD required that building retrofits meet minimum national thermal standards if their floor area exceeds 1,000 m^2—which excludes most housing projects. One of the key changes introduced in 2008 was the elimination of the 1,000 m^2 threshold, implying that all existing buildings undergoing major retrofits would have to meet minimum thermal efficiency levels. It was also suggested that all member states should set target percentages for a minimum percentage of the existing buildings to become energy neutral in 2015 and 2020, and that smart meters to be installed in all new buildings and retrofits.

However, after long negotiations with all member states in 2009, it was decided that the recast of the EPBD would not set obligations or thermal standards for the existing buildings. This was due to the strong resistance from member states that judged the draft proposal as over ambitious and administratively burdensome, especially considering economic challenges and austerity programs. The Directive suggests instead that major retrofits must improve energy efficiency but only if doing so is technically, functionally, and—not surprisingly—economically feasible. For new construction, the consensus was easier to achieve. It was agreed that all new buildings would have to comply with tough energy-performance standards and supply a significant share of their energy requirements from renewables after 2020.

The difficulty in introducing more binding targets for the existing stock in the recast of the EPBD illustrates that there is still a question of the extent to which the EU should intervene in member states' legislation on housing and energy, which have hitherto been purely national policy areas. EU Directives would seem to be an effective catalyst for national action on building regulation, but while they may produce new administrative initiatives, legislation based on Directives will not necessarily be effective in reducing environmental load. The energy performance of the current housing stock varies among EU countries, as do climate and economic conditions, and the member states are at different stages regarding the energy performance of their housing stocks, making uniform requirements across member states difficult. It seems to be left open for each member state to decide whether to adopt mandatory policies (as in Germany) or leave it to the market and stimulate efficiency with fiscal incentives (as in the UK). This leads to a discussion of issues involved in regulating for energy efficiency in the existing stock, and why this may be unpopular in many countries.

3.3 Regulations Versus Market Led Approach

Direct regulation means policy instruments that seek to impose environmentally benign behavior by imposing legal standards. In EU countries, thermal regulations for new builds have been tightened in recent years to support energy-saving strategies, and revised to conform to EU requirements, such as the EPBD described above. Environmental legislation related to the building industry in EU countries tends to focus primarily on energy, indoor air quality, waste, and emissions of hazardous substances. Direct regulation of energy use in the existing buildings has been initiated relatively recently, and only in some EU member states. In 2002, England and Wales imposed minimum insulation levels for replacement of windows and doors, and efficiency standards for boilers, in the existing buildings, controlled by means of self- certification schemes. Germany first required thermal upgrade measures to be included in retrofits in the same year, but these were unique in that they also covered wall, roof, and ground floor insulation (as outlined in Chap. 2). Some countries like Finland are now in the process of introducing mandatory thermal regulations for those retrofits that require building permission, although exceptions can be made depending on the function of the building, its heritage status—and economic viability of thermal measures (YM 2013).

Direct regulation on the existing buildings can operate by means of standards for singular measures, such as minimum insulation levels for building components, or standards for general energy performance of a building, as now required by the EPBD. In most EU countries, minimum insulation levels in new builds were the first type of energy regulation, introduced in the 1970s. These were gradually transformed into integrative approaches, in which the overall energy demand or consumption of buildings was calculated (Beerepoot 2002). Whole building performance standards, as distinct from standards for individual components, can overcome some of the disadvantages of direct regulation as they enable planners to choose the most economically efficient combination of measures to meet the energy performance goal.

Regulations appear to be effective where there are market failures, such as insufficient demand for energy improvements, a constant shortage of affordable housing in high demand areas, and the split incentive dilemma where a landlord invests in energy upgrades and a tenant reaps the benefits. However, the disadvantages of direct regulation, especially regarding existing housing, include: high administrative costs (if the compliance with the regulations is actually inspected or enforced); possible tolerance of noncompliance by local governments as a result of this administrative burden; and economic implications of trying to force costly improvements on private households, especially when housing is considered to be a basic human right.

These issues might lie behind the German Government's choice not to inspect thermal retrofits or enforce the implementation of the EnEV in practice. The UK Government will face these ethical issues in its ambitious and straightforward-sounding plan to make renting EPC 'E' level property, with the existing occupants, illegal in 2018. Further, it is often considered that as a result of a regulatory approach, innovation will be limited as there are no incentives for performance that exceeds the regulations. Germany has addressed this issue with its KfW subsidy program, which applies only to projects that do exceed the regulations.

The question of control is very important, as homeowners do not usually require building permission to carry out retrofit activities in their own property, unless it is listed or in a heritage area. The issue of control for the existing buildings arises in Directive 2002/91/EC, which demands that EPCs are mandatory. However, the energy certificate only proposes energy efficiency improvements and does not require their implementation. Even so, in Denmark, for example, where energy certificates for buildings were first introduced in 1979 and later considered as a prototype for EPCs, only 50–60% of potential buildings had been registered in the scheme by 2005, despite EPCs being mandatory and not requiring any improvement actions (Beerepoot and Sunikka 2005). There were no inspections and no sanctions for noncompliance. A policy assessment saw the cost-effectiveness of labeling of buildings very poor, and no significant difference between houses with or without the energy label (Togeby et al. 2009).

German thermal retrofit policy stands out from that of other European countries by counting on regulation instead of a free market. This could be due to a different political culture, education, the influence of the Green Party (which was a governing coalition partner when the EnEV was devised), or a long established environmental policy. The first Environmental Programme in Germany dates back to 1971, and in 1994 the principle of sustainable development was laid down in the German Constitution in terms of 'bearing responsibility for future generations'. The National Climate Protection Programme (NCPP) identified the retrofit of existing buildings as a priority (BMU 2000).

Apart from new-build regulations, most EU countries have adopted voluntary and market-led policies for building energy efficiency, relying heavily on the environmental conscience of market parties. Policy in the Netherlands and Finland falls between the two extremes of Germany (regulatory) and the UK (market-led), using very few regulatory measures but involving strong leadership from the government. In the Netherlands the government can influence practice, for example through subsidy criteria in the social housing sector, or draw market parties into voluntary agreements. Predicting the impact of a market-led policy can be difficult, and so far there is little evidence of these approaches' success in achieving policy targets. Leaving the encouragement of thermal retrofits to market parties is also an easy option for the government with little political risk.

This leads to a discussion of fiscal policy instruments. These can be both positive (e.g. subsidies) and negative (e.g. taxes), and can be used both in a market-led and a regulatory approach to support thermal retrofits policies.

3.4 Economic Instruments

Financial incentives are an alternative to command-and-control policy instruments such as thermal regulations. Economic instruments in general aim to influence the economic attractiveness of environmentally benign behavior (both investment-related and operational) and, because the environment can be considered a public good for which insufficient market demand exists, they try to overcome market imperfections (Beerepoot and Sunikka 2005).

3.4.1 Negative Fiscal Instruments

Taxes are often assumed to be the least cost policy instrument to encourage energy efficiency and to provide continuous incentive to the industry's search for more cost-effective technologies (Hasegawa 2002; Siebert 1995). Higher taxes on energy may seem relatively effective in reducing a household's consumption and they also reduce the payback times of energy investments. Environmental taxes that aim to shift taxes away from labor and onto the environment have been implemented, to some extent, in several European countries (Andersen 1994; NOVEM 2002; Sunikka 2003). In Germany, the Ecological Tax Reform was introduced in 1999 to encourage energy saving and promote renewable energy sources, for which a portion of the revenue is used (IEA 2000). This supports the objective of German environmental policy to internalize the external costs of environmental protection: applying the polluter-pays principle would thus require energy-related costs to be fully integrated in home owners' and occupiers' expenses.

In 1996, the Federal Environment Agency in Germany studied energy-induced environmental damage related to habitation, and concluded that energy-related costs amounted to €7.7 billion per year, or roughly €2.5 per m^2 of the housing stock per year (Lintz 2000). However, with the current market price of CO_2 abatement at €7 per tonne, paying for one's CO_2 emissions would have added only 1.4% to the cost of one's heating energy bill, assuming CO_2 emissions of 0.0002 tonnes/kWh of heating energy consumed, and a current retail cost of heating energy of €0.10/kWh. The customer was therefore already paying €500 for every tonne of CO_2 emitted, so an extra €7 would have increased this by 1.4%.

Energy taxes are unpopular with the electorate in general and with industry in particular (Beerepoot and Sunikka 2005). Sinn (2008) maintains that Germany's environmental taxes do not cause any net reductions in GHG emissions but merely export these to less environmentally conscious countries. By reducing demand for fossil fuels in Germany, he argues, they reduce the marginal price on the global market, enabling other countries to buy more. To be environmentally effective, he suggests, negative financial instruments need to be imposed in all countries simultaneously, but in practice this is very difficult.

A further problem with using taxes as a policy instrument is that in order to create more sustainable practice and behavioral change, the price incentive needs to be relatively high. But the total environmental costs for industry and households, including both abatement costs and tax payments, are also likely to be high, and this may induce the government to set the tax at a level too low to be effective. The Regulator Energy Tax (REB), for example, was applied to Dutch households in 2001 (and abolished in 2004). Although this increased energy bills by a third, research shows that only half of the population was aware of it, and only 2% took it into account in their electricity use (Van der Waals 2001).

Further, the tax was partly used to subsidise Energy Performance Advice (EPA), introduced in 2000, which led to an increase in the number of energy-labeled houses. However, 75% of EPA assessments were on rented houses, typically owned by professional housing associations rather than private owners. A contradiction existed in that the approach aimed to perform assessments at 'natural moments', such as when a dwelling was being renovated or a central-heating boiler being replaced (Jeeninga et al. 2001). Hence, the EPA subsidy was often used for investments that would have been made anyway. The average cost of the EPA program was 300 euros per tonne of CO_2 reduction, over 20 times the commercial rate. The administrative costs were high, but applications led to relatively small energy savings, with almost three-quarters of customers indicating that the EPA advice had not changed their planned investments in the energy performance of their home (Harmelink et al. 2005).

The question remains as to how taxation on energy can be increased without penalising low-income households, often in energy inefficient housing, with higher energy prices. Despite assistance from government programs such as the Affordable Warmth Obligation in the UK, these households have fewer financial resources to invest in energy-saving measures. It has been argued, therefore, that imposing energy taxes on households causes greater inequality between rich and poor (Anker-Nilssen 2003), so heavy taxation of end-user energy may be neither an advisable nor politically viable option.

3.4.2 Positive Fiscal Instruments

A number of European countries have introduced financial incentives, of various kinds, for improving energy efficiency in buildings (de T'Serclaes 2007; IEA 2008; NOVEM 2002; Sunikka 2003). In most cases, the use of subsidies as a policy instrument aims to create market transformation for energy efficient improvements (see Gillich and Sunikka-Blank 2013).

The French Government's 2007 environmental and energy sustainability initiative, *Grenelle de l'environnement,* spawned a policy for improving the energy efficiency of buildings, the *Plan Bâtiment.* This is a consensus commitment of government and industry to reduce the energy consumption of buildings by 38% by 2020. It includes demonstration projects, and offers an interest-free loan of up to €30,000 for energy upgrades on existing homes, the *Éco-prêt à taux zéro.* Loans have to be repaid within 10 years (principle only) and cover thermal upgrade measures such as insulation, new windows, heat pumps and heat recovery ventilation systems, plus labor, and on-site supervision (Ecocitoyens 2012). Further, under the 2009 Finance Law interest-free loans are offered for the purchase of new or existing homes, with the size of the loan increased if the thermal quality of the home exceeds current building code requirements. Also, tax credits for interest paid on home loans have been modified to incentivise high thermal standards (Pasquier and Saussay 2012).

In 2012, as promised by incoming President Hollande, the French Government announced its aim to encourage an annual retrofit of one million homes. The costs of these retrofit projects will partly be financed by income from the auctioning of CO_2 allowances under the Emissions Trading System (ETS). Further, the French Government wants to introduce progressive energy tariffs aimed at reducing utility bills for energy efficient households, and extend lower social rates for natural gas and electricity for four million low-income households. The costs of a more complex tariff system would need to be met by energy providers.

A strategy for making positive fiscal instruments more effective, and their effects more measurable, is to connect them to the energy reduction targets of voluntary agreements. This policy has been adopted for example in relation to the social housing sector in the Netherlands (see Sunikka and Boon 2004). In Finland's National Climate Strategy and its associated Energy Conservation Programme, voluntary energy conservation agreements play a central role in the implementation of energy policy. Energy conservation agreements are framework agreements made between the Ministry of Trade and Industry (KTM) and various sector organizations, including the real estate sector, though excluding private households. The agreements are voluntary but financially supported and monitored by the government. Participating companies receive 40% financial support for energy audit costs, 15–20% for energy improvement investments and for expenses in setting up an environmental management system under international standard ISO 14001, and up to 40% for new technology investments.

Between 1996 and 2003 the government invested 16.1 million euros in the agreements. Heikkilä et al. (2005) estimate a return of five euros for each euro of public investment. In the real-estate sector, the target was to reduce heating energy consumption by 10% by 2005 compared to 1990 levels and 15% by 2010 but this proved to be over ambitious. Between 2000 and 2004, heating energy consumption per square metre of floor area (kWh/m^2) was reduced by 3.4% compared to 2000 levels, and the heat index (kWh/m^3) by 7% (Hekkilä et al. 2005). Average electricity consumption had stabilized in 2004 to the 2000 level, and the aim was to turn the trend to decline in 2005.

Nevertheless, financial support in the estate sector was mostly given for energy audits, whereas the implementation of the improvements suggested by the audits has been very slow. Again this reflects the problem of moving from energy certificates to actually implementing the recommended measures. Another problem is labor intensity: the real-estate sector required the largest rate of investment to energy efficiency gain: the expenses of the scheme compared to the energy saved were estimated to be around €37.60 per MWh/a.

A difficulty with subsidized voluntary incentive programs is that they can lead to the 'free rider effect', where recipients receive funding for upgrades they would have done anyway. For example, in 1978 the Dutch Government established a large investment subsidy program, the National Insulation Programme (NIP), for improving energy efficiency in the existing housing stock. Kemp (1995) showed that there was only a weak positive relationship between the subsidy for thermal home improvement and the diffusion of thermal insulation technologies. The program mainly provided receivers with a 'windfall gain', helping them in the direction they were already planning to take (Beumer et al. 1993). This effect has been found with other environmental subsidies (Tweede Kamer der Staten Generaal 1987; Vermeulen 1992).

Germany's Federal subsidy system is also based on voluntary thermal improvements, but only for those that go beyond the legal requirements (see Chap. 2). This has the effect that only the least economically efficient aspects of thermal upgrades are being subsidized. The marginal cost of upgrading a building to a standard 20% better than the minimum thermal standard can be in the order of €0.20/kWh of energy saved over the technical lifetime of the retrofit measures, equivalent to €825 per tonne of CO_2 saved (Beecken and Schulze 2011).

An essential question related to both regulatory and market-led retrofit policy approach is: what kind of thermal retrofit measures should be demanded and/or subsidized? The following section explores this issue.

3.5 Deep Versus Incremental Retrofits

A question frequently asked regarding thermal retrofits is should we go deeper, or should we go broader? Deep retrofits are technically demanding and expensive, not only in absolute terms but also in euros spent per kWh of energy saved. Shallower or incremental retrofits are less difficult technically, cheaper to implement, save more energy per euro invested, and can be rolled out quickly to a larger constituency. Despite the common European Directives (Sect. 3.2), retrofit strategies can differ between European countries not only in their implementation strategy (see Sects. 3.3 and 3.4) but also by the choice and depth of technical measures that are encouraged.

Germany's thermal retrofit policy can be described as a deep retrofit approach. The commitment to deep cuts in household heating energy is reinforced by the policy target of 80% GHG emission reduction by 2050, which has been directly translated into the building sector (see Chap. 2). The limitations of this approach are beginning to be recognized. Increasing criticism from homeowners and the building industry is evidenced in recent television reports and a spate of press articles. Academic criticism has come from recent peer-reviewed articles in Germany's highly technical building physics periodical, *Bauphysik* (Beecken and Schulze 2011; Greller et al. 2010; Schröder et al. 2010, 2011). Meanwhile a government-commissioned report authored by some of Germany's leading building physicists concluded that a further tightening of thermal standards would violate the economic viability criterion (Hauser et al. 2012). Despite Germany's ambitious CO_2-reduction goals and high confidence in its technical ability, the German Government has now recognized that its thermal policies for both new builds and retrofits were going too far too fast. The plan to tighten new build thermal standards by a further 30% in 2012 has been replaced by a 12.5% tightening now scheduled for 2014 and possibly a further 12.5% in 2016. The plan to tighten thermal retrofit standards by 30% in 2012 has been shelved, with no timetable for future updates. Ironically, Germany is again proving to be a world leader: while others such as the UK remain committed to making all new homes 'nearly zero energy' by 2016, Germany has effectively stepped back from this goal.

The main reason for this slowdown is that the depth of thermal quality required by the EnEV is proving too difficult to achieve economically for new builds, and too difficult both technically and economically for retrofits. This makes new homes unaffordable and dissuades homeowners from retrofitting, thus reducing national fuel savings. Meanwhile, many homeowners simply ignore the regulations and retrofit to whatever standard suits them (Galvin 2012). It also pushes the cost of saving CO_2, through retrofits, to hundreds of euros per tonne, compared with €10–€30 per tonne for energy-efficiency upgrades in other sectors (Sinn 2008).

In Finland, a similar strategy of deep retrofits and high insulation levels has been encouraged, and the retrofit of state-subsidized housing aims to bring dwellings up to new build standards. Aside from difficulties of economic viability (cf. Chap. 4), research indicates that complaints of mould and poor indoor air quality can be associated with increased insulation and insufficient ventilation. This is partly due to the speed of tightening of the thermal regulations, without considering whether the building industry has been able to adapt its standard construction solutions to changes in building physics that occur as a result of excessive extra insulation (Vinha 2011).

Similar complaints, though to a lesser extent, are heard in Germany, where natural ventilation accounts for nearly 100% of the residential stock. In Finland, France and the Netherlands the share is approximately 30, 40 and 60%, respectively (Meijer et al. 2009).

In France, by contrast, the trend has been toward small retrofits and improvements requested by tenants, supported by the holistic High Environmental Quality (*Haute Qualité Environnementale*) concept (Sunikka 2001). In the Netherlands, the focus of sustainable retrofits has been more on seeing energy efficiency improvements in the context of comprehensive urban renewal projects, including social sustainability targets, although this has recently changed (see e.g. Sunikka and Boon 2004). A number of local schemes in the UK also focus on small incremental upgrades, such as free loft insulation and filling cavity walls.

In theory, we can only reach a national 80% reduction goal by deeply retrofitting every home by an average of 80% savings. While some therefore argue that deep retrofits are the only way forward, others point out that in countries where only deep retrofits are permitted, the rate of retrofits is very low and the actual savings achieved per retrofit are far less than 80% (cf. Tuominen et al. 2012).

But if technical retrofit measures can only partly improve the energy efficiency of the existing housing stock, a further option to consider is behavioral change. Some key elements of this issue are introduced in the next section.

3.6 Technical Measures Versus Behaviour

Occupant behavior and household characteristics are important contributors to domestic energy consumption (Faber and Schroten 2012; Guerra-Santin 2010). Heating energy savings achieved through retrofit measures can be remarkably lower than calculated (Haas and Biermayr 2000), often less than 50% of the expected savings (Simons 2012). Yet current policies in European countries, including Germany, have yet to harness the potential of user behavior to energy savings.

Insights into the effects of behavior on energy use were illustrated in a recent UK study. Sunikka-Blank et al. (2012) describe a thermal retrofit in a Council housing estate in Trumpington, Cambridge. Prior to the retrofit the home was monitored using data loggers and comfort surveys. Temperature measurements were taken in all rooms, at 10-min intervals for 24 h during a 7-day period in March, when the outdoor temperature averaged approximately 2.5 °C. The occupants were an unemployed couple and their three sons. They were using pre-pay meters for their energy costs, as their economic situation disqualified them from contract schemes. This made their heating energy more expensive than for most householders.

Although the couple spent most of the time at home, they said they turned the heating on in winter for only 5.5 h a day at 20 °C. The recordings showed large indoor temperature variations, tracking 10–15 °C above the outdoor temperature, with evening peaks of 25 °C in bedrooms and early morning troughs of 16 °C in the hallway and kitchen. Questionnaires, completed by the occupants three times a day, showed that the children had less tolerance of temperature change, had the highest mean comfort temperature, and wore less clothing than the parents at home.

In order to better understand the characteristics of the case study household compared to other households in the area, a behavioral survey was conducted in 13 identical Council owned properties. Over 39 individuals in 13 households responded to the questionnaire on heating, the use of appliances, occupancy time, level of clothing and heating habits. Households' reported thermostat settings ranged from 16 to 25 °C, the case study being toward the higher end of this scale. The number of heating hours also showed a wide range, from 5 to 24 h per day. Further, the household with the highest energy demand in the survey heated their house more than three times longer than the family with the lowest energy demand.

While it is not possible to generalize behavioral patterns from such a small sample, the findings concur with those of other recent studies. In a study in Denmark, Gram-Hanssen (2010) found households using three or more times as much energy for heating as their neighbors living in identical homes. In studies in the Netherlands, Guerra Santin (2010) found that the quality of building construction plays only a limited role in determining actual energy performance in domestic buildings. As we show in Chap. 5, it is not unusual to find differences of 600% in energy consumption in homes with identical energy ratings.

The large range of differences in energy consumption habits in similar houses suggests two things. First, it is very hard to estimate standard energy consumption for even identical buildings, in simulations related to retrofit. Lifestyle changes have been found to be more effective in saving energy than increasing the thickness of thermal insulation (Shimoda et al. 2003). Information about the thermal properties of buildings can be a very poor predictor of actual consumption in homes. There is also the possibility that occupants may opt for higher indoor temperatures after a building's thermal properties are improved, resulting in significantly less energy saving than expected.

Second, retrofits should allow sufficient deviation in comfort temperatures, more than conventional comfort theory may indicate. People tend to be more tolerant in their comfort zones if they know they can control the temperature and ventilation, so retrofit strategies should offer adaptive opportunities and not be over-engineered.

In the Trumpington case study, despite being a fuel-poor household and pre-pay meter clients, who were paying more than average for their energy, the tenants' energy use was characterized by high indoor temperatures (reaching 25° during the heating season) and a significant number of entertainment appliances. Every member of the family had a TV set; children had their own mini-fridge. Despite economic constraints, there appeared to be an acceptance of energy wasting behavior. In the case study area, households topped up their meters with around £20 weekly during the heating season, regardless of shifts in energy prices. These actions do not reflect 'rational' economic behavior of seeking the lowest price for the most benefit. 12% of all the UK energy customers (mostly low-income or in debt) are on pre-pay meters and pay higher energy prices than if they were on a contract. Data from Brutscher (2010) indicate that such households like to make frequent and small top-ups regardless of the increased energy tariffs or income level, possibly due to liquidity constraints and loss aversion.

UK policy instruments such as Smart Meters are also based on the rational choice models that assume people make decisions based on rational processing of information. In practice, however, energy behavior does not seem to fit this model. In the Trumpington case study, for example, the occupants did not engage with the smart meter provided. Hargreaves et al. (2010), Shove (2003) and Gram-Hanssen (2010) are pioneering research on household energy behavior that seeks to explain people's choices in terms of routinized practices and habits, rather than rational, economically based decisions. Galvin (2013) applies this approach to a study of household ventilation practices in Aachen, Germany, again finding that indoor routines and habits dominate over economically 'rational' behavior. Uitdenbogerd et al. (2007) also suggest energy relevant household behavior seems to be of habitual character rather than based on economic rationality. This can limit the effectiveness of economic instruments (Sect. 3.4), and leads us to explore quite different approaches to changing energy use behavior.

3.7 Conclusions and Implications

The EU member states have agreed on a 20% energy efficiency target by 2020. The main policy instruments of the EU to stimulate thermal retrofits are the recast Energy Performance of Buildings Directive (EPBD) and the Energy Efficiency Directive (EED), although both Directives were weakened in the negotiation stages. As a part of the EED, each member state is required to draw up a roadmap to make the building sector more energy efficient by 2050. Nevertheless, Germany has halted its drive to continually tighten thermal retrofit regulations, and delayed and reduced its next step in tightening new build thermal standards. Our analysis (see Chap. 4) suggests it will fall far short of the 20% target, in the residential building sector, by 2020.

Most EU countries have to a large extent adopted voluntary and market-led policies for promoting thermal retrofits, relying heavily on the environmental conscience of market parties. Where fiscal incentives are used, taxes are often seen as the least cost policy instrument, but high energy taxes are unpopular with the electorate in general and with industry in particular, including in Germany, and it is difficult to impose a level high enough to bring desired results without penalising lower income households. Subsidies are used to promote thermal retrofits, for example the interest-free tax in France and KfW subsidies in Germany. Some research indicates a free-rider effect and a weak positive relationship between subsidies and the diffusion of thermal insulation technologies, but free loft insulation in the UK does seem to be reaching households that would otherwise not insulate their homes.

German belief in technical fixes and deep retrofits could be rooted in the ambitious policy target of an 80% reduction by 2050, and the fact that most post-war apartment housing has solid, plain, thermally inefficient walls that lend themselves to external wall insulation. Ironically, however, this sector is one of the slowest to undertake retrofitting. There is also a backlash against deep retrofit regulations in Germany among homeowners, the building industry and some academics and research institutes, for reasons ranging from excessive marginal costs and technical difficulties, to complaints of indoor mould and discomfort after retrofitting. Excessive insulation levels in countries like Finland have also brought complaints of mould and indoor quality risks.

EU countries appear to share a consensus on the energy-focused concept of thermal retrofit and carbon emissions targets, usually by 60–80% by 2050, for example in the Netherlands, Germany, and the UK (MVROM 1999; BMU 2000; DTI 2003). The strongest driving forces to make vague aims more specific have been the Kyoto Protocol and the European Union Directives, which put pressure on governments to achieve measurable energy savings. Quantitative targets, however, are still set over a very long time frame and the measures required to achieve them are not necessary defined. Some policy targets seem to be overoptimistic, with policymakers tending to overestimate outcomes when under pressure to deliver overall targets (Wagner et al. 2005). There may need to be a rethink of how to reach goals of 60–80% reduction in home heating energy consumption by 2050.

An avenue yet to be systematically explored is the potential energy savings in household behaviour change. The EU Commission is now promoting this as a significant source of large, untapped energy savings (Faber and Schroten 2012). The case study of low-income households in Trumpington illustrates the large range of heating energy consumption patterns among apparently similar households in identical dwellings. As in many other studies, these consumers' energy choices do not seem to be based on classical models of economic rationality, a possible reason why policy instruments based on such models are failing to bring effective change. Research in this area is moving beyond rational choice theory to explore how habitual and routinized behavior locks consumers into high consumption patterns. If energy efficient, exemplar household routines can be identified, they may then be able to be taught to or copied by others.

References

Andersen MS (1994) Governance by green taxes. Manchester University Press, Manchester

Anker-Nilssen P (2003) Social obstacles in curbing residential energy demand. In: Attali S, Mettreau E, Prone M, Tillerson K (eds) Proceedings of the ECEEE 2003 summer study, St-Raphael, 27 June 2003. European Council for Energy Efficient Economy, Stockholm

Beecken C, Schulze S (2011) Energieeffizienz von Wohngebäuden: Energieverbräuche und Investitionskosten energetischer Gebäudestandards. Bauphysik 33(6):338–344

Beerepoot M (2002) Energy regulations for new building, in search of harmonisation in the European Union. Delft University Press, Delft

Beerepoot M, Sunikka M (2005) The contribution of the EC energy certificate in improving sustainability of the housing stock. Environ Plan B 32(1):21–31

Beumer L, Van der Giessen EC, Olieman R, Otten GR (1993) Evaluatie van de isolatieregeling (SES 1991) en de ketelregeling (SNEV). NEI, Rotterdam

BMU (Federal Ministry for Environment and Reactor Safety) (2000) Germany's national climate protection programme, summary. BMU, Berlin

Brutscher PB (2010) Pay-and-pay-and-pay-as-you-go: An exploratory study into pre-pay metering. In: Presentation at the Cambridge University energy and environment seminar, 5 May 2010

de T'Serclaes P (2007) Financing energy efficient homes: existing policy responses to financial barriers, IEA information paper. IEA, Paris

DTI (Department of Trade and Industry) (2003) Our energy future—creating a low-carbon economy, Energy White Paper. DTI/The Stationery Office, London

EC (2003) Council directive 2002/91/EC of 16 December 2012 on the energy performance of buildings. Off J Eur Commun 65–71

Ecocitoyens (2012) Éco-prêt à taux zero. http://ecocitoyens.ademe.fr/financer-mon-projet/renovation/eco-pret-a-taux-zero. Accessed 18 Oct 2012

Faber J, Schroten A (2012) Behavioural climate change mitigation options and their appropriate inclusion in quantitative longer term policy scenarios—main report. CE Delft, Delft. http://ec.europa.eu/clima/policies/roadmap/docs/main_report_en.pdf. Accessed 19 Dec 2012

Galvin R (2012) German Federal policy on thermal renovation of existing homes: a policy evaluation. Sustain Cities Soc 4:58–66

Galvin R (2013) Impediments to energy-efficient ventilation in German dwellings: a case study in Aachen. Energy Build 56:32–40

Gillich A, Sunikka-Blank M (2013) Barriers to domestic energy efficiency—an evaluation of retrofit policies and market transformation strategies. In: Proceedings of the ECEEE 2013 summer study, Presqu'île de Giens, Toulon/Hyères, 3–8 June 2013. European Council for Energy Efficient Economy, Stockholm

Gram-Hanssen K (2010) Residential heat comfort practices: understanding users. Build Res Inform 38(2):175–186

Greller M, Schröder F, Hundt V, Mundry B, Papert O (2010) Universelle Energiekennzahlen für Deutschland—Teil 2: Verbrauchskennzahlentwicklung nach Baualtersklassen. Bauphysik 32(1):1–6

Guerra Santin O (2010) Actual energy consumption in dwellings, the effect of energy performance regulations and occupant behaviour. IOS Press, Amsterdam

Hargreaves T, Nye M, Burgess J (2010) Making energy visible: a qualitative field study of how householders interact with feedback from smart energy monitors. Energy Policy 38:6111–6119

Harmelink M, Joosen S, Blok K (2005) The theory-based policy evaluation method applied to the ex-post evaluation of climate change policies in the built environment in the Netherlands. In: Proceedings of the ECEEE 2005 summer study, energy savings: what works & who delivers? Mandelieu La Napoule, 30 May–4 June 2005. ECEEE, Stockholm

Haas R, Biermayr P (2000) The rebound effect for space heating Empirical evidence from Austria. Energy Policy 28(6–7):403–410

Hasegawa T (2002) Policies for environmentally sustainable buildings, OECD report ENV/EPOC/WPNEP (2002)5. OECD, Paris

Hauser G, Maas A, Erhorn H, de Boer J, Oschatz B, Schiller H (2012) Untersuchung zur weiteren Verschärfung der energetischen Anforderungen an Gebäude mit der EnEV 2012—Anforderungsmethodik, Regelwerk und Wirtschaftlichkeit: BMVBS-Online-Publikation, Nr. 05/2012. BMVS, Berlin

Heikkilä I, Pekkonen J, Reinikainen E, Halme K, Lemola, T (2005) Energiasopimusten kokonaisarviointi. Ministry of Trade and Industry (KTM), Helsinki

House of Commons (2011) Energy and climate change committee—fuel poverty in the private rented and off-grid sectors: written evidence submitted by the Department of Energy and Climate Change (DECC). DECC, London. http://www.publications.parliament.uk/pa/cm201012/cmselect/cmenergy/1744i_ii/1744vw02.htm. Accessed 18 Oct 2012

IEA (International Energy Agency) (2000) Country analysis briefs, Germany, France and the United Kingdom. IEA, Paris. www.iea.doe.gov/emeu/cabs. Accessed 19 Dec 2000

IEA (International Energy Agency) (2008) Promoting energy efficiency investments: case studies in the residential sector. IEA, Paris

Jeeninga H, Beeldman M, Boonekamp PGM (2001) EPAWoningen—Nadere invulling van de EPA doelstelling voor woningen. ECN, Petten

Kemp R (1995) Environmental policy and technical change: a comparison of the technological impact of policy instruments. University of Limburg, Maastricht

Lintz G (2000) Environmental costs of the construction and the use of residential buildings in Germany. In: Proceedings of the sustainable building 2000 conference, Maastricht, 22–25 Oct 2000. Aenas, Best

Meijer F, Itard L, Sunikka-Blank M (2009) Comparing European residential building stocks: performance, renovation and policy opportunities. Build Res Inform 37(5):533–551

MVROM (Ministerie van Volkshuisvesting, Ruimtelijke Ordening en Milieubeheer) (1999) Beleidsprogramma Duurzaam Bouwen 2000–2004; Duurzaam Verankeren. MVROM, The Hague

NOVEM (2002) Operating Space for European Sustainable Building Policies. In: Report of the Pan European conference of the Ministers of Housing addressing sustainable building, Genvalle, Belgium, 27–28 June 2002. NOVEM, Utrecht

Pasquier S, Saussay A (2012) Progress implementing the IEA 25 energy efficiency policy recommendations. International Energy Agency (IEA), Paris. http://www.iea.org/publications/insights/progress_implementing_25_ee_recommendations.pdf. Accessed 15 Oct 2012

Rosenow J (2011) Different paths of change: the case of domestic energy efficiency policy in Britain and Germany. In: Proceedings of the ECEEE summer study 2011. Belambra Presqu'île de Giens, 6–11 June. ECEEE, Stockholm

Schröder F, Engler HJ, Boegelein T, Ohlwärter C (2010) Spezifischer Heizenergieverbrauch und Temperaturverteilungen in Mehrfamilienhäusern—Rückwirkung des Sanierungsstandes auf den Heizenergieverbrauch. HLH 61(11):22–25. http://www.brunata-metrona.de/fileadmin/Downloads/Muenchen/HLH_11-2010.pdf. Accessed 8 Dec 2011

Schröder F, Altendorf L, Greller M, Boegelein T (2011) Universelle Energiekennzahlen für Deutschland: Teil 4: Spezifischer Heizenergieverbrauch kleiner Wohnhäuser und Verbrauchshochrechnung für den Gesamtwohnungsbestand. Bauphysik 33(4):243–253

Shimoda Y, Fujii T, Morikawa T, Mizuno M (2003) Development of residential energy end-use simulation model at a city scale. In: Proceedings of the eighth international IBPSA conference, Eindhoven, 11–14 Aug 2003

Shove E (2003) Comfort, cleanliness and convenience: the social organization of normality. Berg, Oxford

Siebert H (1995) Economics of the environment, theory and policy. Springer, Berlin

Simons H (2012) Energetische Sanierung von Ein- und Zweifamilienhäusern Energetischer Zustand, Sanierungsfortschritte und politische Instrumente, Bericht im Auftrag des Verbandes der Privaten Bausparkassen e.V. Empirica, Berlin

Sinn HW (2008) Das Grüne Paradoxon: Plädoyer für eine illusionsfreie Klimapolitik. Econ Verlag, Berlin

Sunikka M (2001) Policies and regulations for sustainable building, a comparative analysis of five European countries. Delft University Press, Delft

Sunikka M (2003) Fiscal instruments in sustainable housing policies in the EU and the accession countries. Eur Environ 13(4):227–239

Sunikka M (2006) Policies for improving energy efficiency if the European housing stock. IOS Press, Amsterdam

Sunikka M, Boon C (2004) Introduction to sustainable urban renewal. Delft University Press, Delft

Sunikka-Blank M, Chen J, Dantsiou D, Britnell J (2012) Improving energy efficiency of social housing areas—a case study of a retrofit achieving an 'A' energy performance rating in the UK. Eur Plan Stud 20(1):133–147

Togeby M, Dyhr-Mikkelsen K, Larsen A, Juel Hansen M, Bach P (2009) Danish energy efficiency policy: revisited and future improvements. In: Proceedings of the ECEEE 2009 summer study. Act! innovate! deliver! reducing energy demand sustainably, La Colle sur Loup, 1–6 June 2009. ECEEE, Stockholm

Tuominen P, Klobut K, Tolman A, Adjei A, de Best-Waldhober M (2012) Energy savings potential in buildings and overcoming market barriers in member states of the European Union. Energy Build 51:48–55

Tweede Kamer der Staten Generaal (1987) EvaluatieWIR-milieutoeslag. Tweede Kamer, vergaderjaar 1986–1987, 19858, No. 4, The Hague

Uitdenbogerd D, Egmond C, Jonkers R, Kok G (2007) Energy-related intervention success factors: a literature review. In: Proceedings of the ECEEE 2007 summer study: saving energy—just do it! La Colle sur Loupe, France, 4–9 June 2007. ECEEE, Stockholm

Van der Waals J (2001) CO$_2$-reduction in housing: experiences in building and urban renewal, projects in the Netherlands. Rozenberg, Utrecht

Vermeulen W (1992) De vervuiler betaald: Onderzoek naar de werking van subsidies op vier deelterreinen van het milieubeleid. Rijksuniversiteit Utrecht, Utrecht

Vinha J (2011) Yhteenveto Frame-projektin tuloksista. Presentation at: Future envelope assemblies and HVAC solutions (FRAME)—tutkimuksen päätösseminaari, Tampere, Finland, 8 Nov 2011. http://www.rakennusteollisuus.fi/frame. Accessed 4 Jan 2013

Wagner O, Lechtenböhmer S, Thomas S (2005) Energy efficiency—political targets and reality. Case study on EE in the residential sector in the German Climate Change Programme.

In: Proceedings of the ECEEE 2005 summer study, energy savings: what works & who delivers? Mandelieu La Napoule, 30 May–4 June 2005. ECEEE, Stockholm

YM (Ympäristöministeriö) (2013) Rakennusten energiatehokkuusdirektiivi. http://www.ymparisto.fi/default.asp?contentid=389101&lan=FI. Accessed 4 Jan 2013

Chapter 4
The Technical Potential and Limitations of Thermal Retrofits in Germany

Abstract Thermal retrofitting of existing homes is widely seen as an economic and technically feasible way to reduce domestic heating energy consumption, which accounts for 14.6% of Germany's total energy consumption. But energy savings from retrofits are dependent on the laws of physics and the geometric and physical characteristics of the actual housing stock. This chapter outlines these issues, first describing the main features and components of thermal renovation in the German context. It then explores the two metrics used to assess thermal performance in homes, showing how one of these departs precariously from the laws of physics, producing unrealistic estimates of how dwellings perform prior to and after a retrofit. The structural, physical characteristics of German homes are then examined alongside the measures available for thermal upgrades, highlighting the challenges typically encountered in trying to bring these homes to a high standard of thermal performance.

Keywords Thermal retrofit · Housing stock · Energy saving · CO_2 reduction · German policy

4.1 Introduction

It is frequently asserted that there is a huge energy saving potential in thermal renovation of existing homes in Germany (BMVBS Bundesministerium für Verkehr and Bau und Stadtentwicklung 2011; DENA Deutsche Energie-Agentur 2010; IEA International Energy Agency 2008). 18% of German energy consumption takes place in homes (BMWi Bundesministerium für Wirtschaft und Technologie 2011:14). In 2007, a typical recent year, 67% of the energy consumed in German homes was for space heating and a further 14% for water heating (BMWi Bundesministerium für Wirtschaft und Technologie 2010:25). Hence, domestic space and water heating accounted for 14.6% of German energy consumption. This can be reduced through insulation, improved air-tightness, and

R. Galvin and M. Sunikka-Blank, *A Critical Appraisal of Germany's Thermal Retrofit Policy*, Green Energy and Technology, DOI: 10.1007/978-1-4471-5367-2_4, © Springer-Verlag London 2013

increased efficiency of boilers. It is common among German policy actors to refer to a graph published by McKinsey and Company (2009), which claims that 'insulation retrofit' of buildings, if 'pursued aggressively', will bring a positive gain of €60 for every ton of CO_2 emissions saved.

But what is the technical potential of heating energy saving in the German housing stock? Closely related to this is the question as to what level of technical upgrade is economically feasible. There is always a trade-off between these two factors: technology and economics. Clearly we can make a dwelling as energy-efficient as we like if we wrap it in a thick enough layer of insulation, plug all the air leaks, and use the best possible technology for the heating and ventilation systems. But because it has to work in the real world, thermal upgrade technology has developed with economy in mind. For clarity of argument presentation we focus in this chapter on technical potential and limitations, referring to economics where this becomes immediately relevant. In Chap. 6 we will focus on economic viability.

Section 4.2 of this chapter gives an overview of the thermal upgrade features that are generally possible and usually undertaken in existing German homes, in the context of the basic principles of heat loss in residential buildings. Section 4.3 delves deeper into the physics of thermal upgrades, and how this effects what is technically possible. Section 4.4 explores some of the practical problems encountered in applying modern thermal upgrade technology to the homes that actually exist in Germany. Section 4.5 offers conclusions, and discusses implications for thermal retrofit policy.

4.2 Thermal Upgrade Features and Components

4.2.1 Heat Loss in Residential Buildings: Conduction, Radiation and Convection

Figure 4.1 is a schematic of a typical German detached house prior to a thermal retrofit. Heat energy is lost through the roof, walls, basement ceiling, windows, and external doors, and there are further losses due to energy inefficiency in the boiler. The heat energy is transmitted to the outside world through conduction, radiation, convection, and air leakage. Heat losses from the boiler occur directly through the chimney or flue.

Aside from boiler inefficiencies and air leaks, indoor heat is transported through the building envelope—the walls, roof, windows, exterior doors, and basement ceiling—by conduction. The rate of conduction (i.e. of heat energy loss) is directly proportional to the difference between indoor and outdoor temperature, and inversely proportional to both the thickness of the relevant portion of the building envelope and its U-value, a metric for the conductivity of a medium (explained in more detail in Sect. 4.3).

Fig. 4.1 Schematic of an existing detached house, not thermally upgraded, showing basic heat loss through the roof, walls, basement ceiling, windows, door and energy inefficiency in the boiler

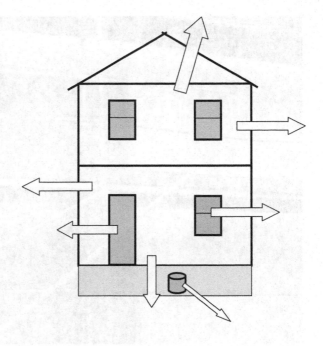

To inhibit conduction of heat, would-be renovators therefore need to reduce the U-values of the various parts of the building envelope. In Germany this is most commonly achieved by attaching a layer of insulating material such as Styropor®, a petroleum-based product similar to polystyrene and developed by the chemical firm BASF, to the walls, roof, and basement ceiling, and replacing the windows and external doors with models that have low U-values (see Figs. 4.2 and 4.3). In most cases wall insulation is glued and screwed to the outside surfaces of the walls and covered with a protective render of fiberglass and plaster. Common alternatives to Styropor are Neopor®, which is a form of Styropor impregnated with graphite to give a reflective effect, thus further inhibiting heat transport, and rock-wool, a fireproof, environmentally friendly alternative made from pulverised stone.

Roof insulation is commonly added during re-tiling of roofs. A layer of physically robust insulating material is laid on top of the rafters, and the new tiles are fitted on top of this. An alternative method of roof insulation, without removing the tiles, is to fit layers of glass wool, rock-wool, hemp, or wood fiber under the tiles, with a first layer inserted between the rafters and another layer covering this.

Internal wall insulation—attaching insulation to the inside of walls—is not popular in Germany because it causes the substance of the wall to cool below the dew-point, bringing ice formation and moisture problems to the inner wall substance where it attaches to the insulation. It can also drastically reduce the living area, as wall insulation has to be 12–16 cm thick to meet legal requirements.

But radiation, not conduction, is the main *ultimate* cause of heat loss. The heat conducted through the walls, windows, and roof warms the outside surface of the

Fig. 4.2 A commonly used insulation material in walls, roof and basement ceiling is Styropor®
(BASF), similar to polystyrene

building, which then radiates this heat into space and to surrounding objects. The
rate of heat loss by radiation depends on the temperature of the outside surface of
the building envelope, the temperature of the surfaces facing it, and the 'emissivity' of these surfaces (the Stefan-Boltzmann law). A so-called 'black body' has
an emissivity of 1.0, meaning it radiates completely, while lighter color surfaces,
and surfaces made of various substances, radiate less intensively.

Painting the walls of a house a light color is therefore a sensible strategy, but
the main factor in reducing heat loss by radiation is to keep the outside surface of
the building envelope as cold as possible. Heat loss is proportional to the difference between the fourth power of the absolute temperature of the surface of the
building envelope and the fourth power of the absolute temperature of the surfaces
facing it ($T_E^4 - T_S^4$) so a small change in surface temperature makes a big difference
to the heat loss. If, for example, the temperature of the surroundings is −5 °C
(268.15 K), a building envelope with a surface temperature of 4.7 °C (277.85 K)
will radiate (i.e. lose) twice as much heat energy per hour as one with a surface
temperature of 0 °C (273.15 K); a building envelope with a surface temperature of
13.6 °C (286.75 K) will radiate 4 times as much; and one with a surface temperature of 20 °C (293.15 K) will radiate 6 times as much.

On a clear winter night the surface facing the roof of a house is the sky, which
may be effectively much colder than −5 °C, thus increasing the heat loss through

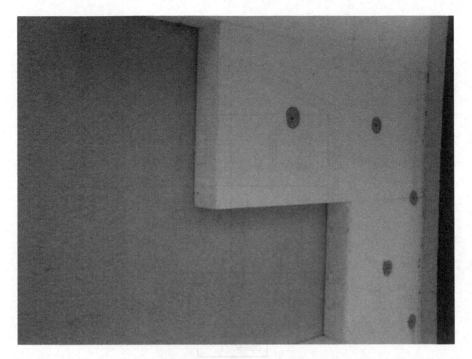

Fig. 4.3 Wall insulation is usually glued and screwed to the outside surfaces of the walls and covered with a render

radiation. This underlines the importance of keeping the outside surface of the roof as cool as possible.

To keep the outside surface of the building envelope cool we have to reduce the rate of conduction of heat through the building envelope. This is achieved, as we noted above, by improving the insulation.

Figure 4.4 displays a typical set of insulation features for a German thermal retrofit on the building shown in Fig. 4.1. To reach the standards in the Energy Saving Regulations *(Energetische Einsparverordnung—EnEV)*, the wall insulation is generally 16 cm thick, with 12 cm on the basement ceiling and 22 cm in the roof. The roof in our schematic has been raised to accommodate this thickness.

There is a further cause of heat loss: convection. Convection simply means that warm air rises and cooler air falls. Therefore, conventional heating systems cause a build up of warm air under the ceiling, where we do not need it, and a dearth of warm air at floor level, where we do need it. They also cause a general drift of warm air upwards in the house, accumulating under the roof. This, coupled with the sky-facing orientation of roofs, makes them a crucial source of heat loss and an essential candidate for generous insulation. For this reason German building regulations require roof insulation to be nearly 1.5 times as thick as insulation on lower storey walls (see Table 4.1 in Sect. 4.3.1).

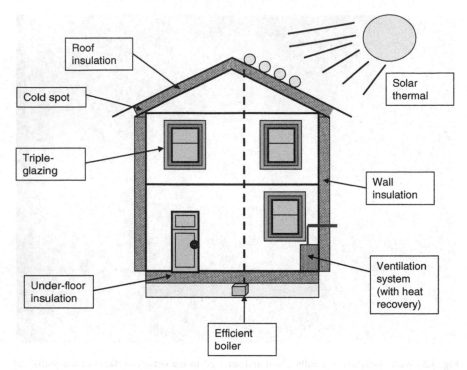

Fig. 4.4 Schematic of an existing detached house showing possible thermal upgrade features

Convection can also add to heat losses from the outside surface of the building. Cool outdoor air in contact with a warm building envelope will heat up and rise, being replaced by further cool air from below, which will also heat up, and continue the process. Falling rain will also draw heat from the building envelope as it runs down the roof, walls and windows.

Table 4.1 Recommended U-values for components of building envelope for new builds and partial thermal retrofits, as given in EnEV (Energieeinsparverordnung) 2009, with examples of insulation media and thickness to achieve these values.

Component	U-value (W/m²K)	Example of medium	λ-value (W/mK)	Thickness required (cm)
Lower wall	0.28	Styropor	0.035	12.5
Wall reaching roof	0.20	Styropor	0.035	17.5
Roof	0.20	Wood fibre	0.040	20
Basement ceiling	0.35	Styropor	0.035	10
Basement wall	0.35	Neopor	0.030	9
Window	1.30	Vacuum/inert gas	0.004	3.1
External door	1.80	Vacuum/inert gas	0.006	3.3

Source EnEV Energieeinsparverordnung (2009) and authors' calculations

4.2.2 Surface Area, Building Geometry and Heat Loss

A further factor in heat loss is the surface area of the building envelope. The greater the outside surface area, the more heat will be lost through radiation and convection. This has three important implications.

First, extra corners, dormers, towers, balconies, chimneys, gargoyles and recesses increase heat loss significantly, because they increase the surface area of the building envelope. Balconies can be a particular problem for thermal quality, as they transport heat by conduction from the wall to which they are attached and offer large surface areas for radiation of heat to the surroundings. Most new homes in Germany have floating balconies, supported by pillars running directly into the ground, with a bare minimum of attachment to the side of the building—usually just two bolts—and an air gap of a few millimeters between the balcony and the wall. This prevents conduction of heat from wall to balcony via so-called 'cold bridges' or 'thermal bridges', effectively reducing the building's surface area considerably. It is not uncommon in thermal renovation in Germany to literally cut balconies off and replace them with a floating model.

Second, the geometric shape of a building serves to determine its ratio of surface area to volume. A shape approximating a cube is more energy-efficient than a cuboid (oblong shape), even if both have the same volume, because a cube has less surface area per unit volume. Theoretically, the ideal building shape for heat energy efficiency is a bland, plain structure; it is more challenging to achieve energy savings targets in oblong or multi-winged buildings than in buildings that approximate a cubic shape. However, among homeowners we have interviewed, some complain that Germany already has warning examples of unattractive houses that are energy efficient but with small windows and no quality of design.

Third, the larger the building, the lower the ratio of surface area to volume, and therefore the more energy-efficient the building is. Doubling the length of a building (and its height and width) increases its surface area by a factor of 4, but its volume by a factor of 8 (since $2^2 = 4$, while $2^3 = 8$), thus effectively halving the ratio of surface area to volume, and therefore halving the heat loss per square meter of living space (assuming floor area is proportional to volume). Semi-detached houses, terraced houses, and apartment blocks attached to one another in a long row function, geometrically, as large buildings. They therefore have the advantage of less surface area exposed to the elements than individual buildings of the same size. Small-detached houses, by contrast, are challenging for heating energy efficiency upgrades as they have so much exposed surface area in comparison to their volume.

4.2.3 Air Leakage and Solar Gain

A further, and very significant, source of heat loss is air leakage. Many older buildings suffer drafts through gaps in the window and door fittings. This can lead to replacement of indoor air with outdoor air several times per hour. In an average-sized German home of 87 m^2 living area and 2.6 m high rooms, about 1 kWh of energy is required to heat the indoor air from 5 °C to 18 °C (Galvin 2013). Air leakage can cause several full exchanges of air per hour, consuming several kWh per hour, leading to an excess annual heating energy consumption of thousands of kWh, costing hundreds of euros.

As the schematic in Fig. 4.2 shows, the windows and external doors have been replaced with low U-value, air-tight models. The air-tightness causes a build up of CO_2 and moisture in the indoor air. Moisture can lead to mould growth, particularly where there is condensation. This occurs on indoor surfaces that are colder than their surroundings, typically on window frames and in corners where the ratio of surface area to volume is high (and hence an excess of heat is being lost through conduction).

One solution to moisture and CO_2 build up is to install a ventilation system throughout the whole dwelling, preferably with a heat recovery unit so that the heat in the outgoing air is transferred to the incoming air. This is technically straightforward for new homes, as it can be designed into the structure of the house. For older homes it can be very expensive, as ducts need to be built throughout the already existing structure. The next best solution is manual ventilation. However, this can lead to substantial energy losses when householders keep windows open too long (Galvin 2013).

Some of these heating energy losses are offset by free heat from the sun, known as 'solar gain'. New builds can optimise this with south-facing windows, but existing homes usually leave little scope for improvement. However, modern window glass is designed to maximise solar gain and minimise heat loss by radiation from indoors.

New homes are now required by regulation to include some form of on-site renewable energy source, and solar thermal collectors are popular because they offer a relatively high return, in energy gained per euro invested. These supplement water heating are integrated with the boiler. Many thermal retrofit projects now include solar thermal collectors if the existing building has an unobstructed south-facing roof. This requires extra plumbing, since the roof is usually a long way from the boiler. It also requires a compatible boiler, so it is best done when the current boiler is nearing the end of its useful life.

4.3 Germany and the Physics of Thermal Upgrades

The thermal performance of German residential buildings is assessed in terms of two different but related measures: transmission losses and quantity of energy consumed. Other important issues in this regard are heating degree-days and the standard methodology for calculating energy consumption.

4.3.1 Transmission Losses

Transmission losses refer to the heat lost by conduction through solid media, such as walls, windows, and roofs. These losses are measured in watts per square meter of the conducting medium, per degree Kelvin difference between its two sides (W/m^2K). For individual components they are given the term 'U-values'. The lower the U-value, the better the insulation medium. The walls of older houses typically have U-values of around 3.0 W/m^2K or higher, while a refurbished wall with 16 cm of Styropor insulation will typically have a U-value of around 0.22 W/m^2K. The U-value of typical modern, double-glazed windows is higher, at 1.3 W/m^2K, since these are considerably thinner than 16 cm of insulation. The average U-value of the entire building envelope is called H_T in Germany. For an old, detached German house H_T is typically around 4.0 W/m^2K, while the maximum permissible H_T value for a new build of the same type is 0.5 W/m^2K, and for a comprehensive retrofit to EnEV standards this would normally be about 0.7 W/m^2K.

Another metric often used in building engineering, but not frequently in Germany, is 'thermal resistance', or 'R-value'. The R-value of a medium is simply the reciprocal of its U-value: $R = 1/U$.

It can be difficult to conceptualize the meaning of U-values (Watts per square meter per degree Kelvin), but perhaps it becomes clearer if we note that this is the same metric as 'kilowatt hours of energy lost every thousand hours, through every square meter of wall area, for every degree difference between indoor and outdoor temperature'. Consider, for example, a typical detached house with 215 m^2 of wall, roof, and ground floor. If the average U-value of this total area (i.e., H_T) is 3.0 W/m^2K and the difference between indoor and outdoor temperature averages 15 °C, it will loose $3.0 \times 215 \times 15 = 9{,}675$ kWh every thousand hours. If the average U-value is reduced to 0.7 W/m^2K, only 2,258 kWh will be lost. On paper this represents a saving of 76%, but in practice further factors reduce these theoretical gains considerably, as we shall see in later chapters.

The thickness of insulation required to reach a particular U-value is worked out from the insulating material's 'conductivity', or 'λ-value' (lambda-value). λ-values have the nonintuitive metric, W/mK, i.e., watts per meter per degree difference in temperature. Again this is easier to conceptualize as 'kilowatt-hours lost per thousand hours, per meter of thickness of insulating material, per degree difference between its two sides'. The lower the λ-value, the better the insulation material. Typical λ-values are: Glass wool $\lambda = 0.040$ W/mK; Styropor® $\lambda = 0.035$ W/mK; Neopor® $\lambda = 0.030$; brick $\lambda = 0.77$ W/mK and solid concrete $\lambda = 1.4$ W/mK.

To find the U-value for a particular thickness of insulation material we simply divide the λ-value by the thickness of the material, in meters. Hence a 16 cm thick block of Styropor has $U = 0.035/0.16 = 0.22$ W/m^2K. Table 4.1 gives the recommended U-values for new builds and partial thermal retrofits, as prescribed in EnEV (Energieeinsparverordnung 2009). As we noted in Chap. 2, these U-values can be varied, provided the building's average U-value (i.e. its H_T) falls under the maximum for the building's geometry (in the range 0.4–0.65 W/m²K), and the theoretical heating fuel consumption (Q_P) is within the limit for an identical building using the recommended U-values.

4.3.2 Quantity of Energy Consumed

We now consider the second measure of thermal performance: quantity of heating energy consumed. This is measured in kilowatt-hours of primary energy consumed per square metre of 'useful area', per year (kWh/m²a) and is given the symbol Q_P in German regulations and literature. 'Useful area' includes the living area inside an apartment's door, plus a proportionate area of stairwell, landing, common loft space, etc. Since the 2002 upgrade in the thermal regulations, new builds in Germany have been required to satisfy both measures of thermal performance: average transmission losses (H_T, i.e. average U-value in W/m²K) and heating energy consumption (Q_P, in kWh/m²a). For retrofits the rules are slightly different. *Partial* retrofits have to satisfy only the first criterion: the U-values for their thermal upgrade measures must meet new build standards. *Comprehensive* thermal retrofits, however, must instead satisfy the second criterion, Q_P, but at a less stringent level: their heating energy consumption can be 40% greater than that for a new build of the same size, geometric shape, and configuration of components. The average permissible Q_P for a comprehensive renovation is 100 kWh/m²a, but for smaller and irregular-shaped buildings it is higher (less stringent but usually harder to achieve), while for larger, more regular-shaped buildings and those adjoining other buildings it can be considerably lower (more stringent but usually easier to achieve).

While H_T is a robustly objective measure, based on the physical properties of insulation materials, Q_P is somewhat artificial. We can get an idea of the approximate relationship between these two metrics by considering the notion of 'heating degree-days'.

4.3.3 Heating Degree-Days

The heating degree days (HDDs) of a city or region are the equivalent number of days (i.e. hours divided by 24) that the outdoor temperature is below an indoor base temperature, multiplied by the average difference between these two temperatures. For the Lindenberg district of Berlin, for example, using a base indoor temperature of 15.5 °C, there were 2,524 degree-days between October 2011 and September 2012 (www.degreedays.net). To obtain an approximate value for the total heating energy consumption for a building in Lindenberg, Berlin that is heated continuously to the base temperature (and no higher), we use the formula:

$C = H_T \times A \times 24 \times HDD/1000$, where A is the total area of the building envelope (not the floor area).

The factor 24 converts days to hours, while the divisor 1000 converts watts to kilowatts, producing an answer in kWh. For a detached house with building envelope area 225 m^2 and $H_T = 2.0$ W/m^{2}K, this is:

$$2.0 \times 225 \times 2,524/1000 = 11,358 \text{ kWh}.$$

If the living area of the house is the German average of 87 m^2, and the 'useful area' is 20% greater than this, a very approximate estimate for Q_P would be:

$$11,358/(87 \times 1.2) = 109 \text{ kWh/m}^2\text{a}.$$

Since both these answers are directly proportional to H_T, reducing H_T by insulating the building envelope reduces Q_P, the theoretical quantity of energy consumed, in proportion to the reduction in H_T. However, using HDDs in this way gives only the crudest idea of the real annual energy consumption, as this also depends on a number of other factors: how high the occupants set the thermostat; for how much of each day and night the heating is on; which of the rooms are heated and at what times; how often and for how long the occupants open which windows and external doors (especially for ventilation); how much extra heat is gained from human activity, cooking, and electrical equipment; and how much heat is won from or lost to adjoining apartments below, above, or next to the one in question.

4.3.4 The German Standard for Calculating Theoretical Heating Consumption

In an attempt to deal with this array of factors, Germany uses a formula, published by the German Standards Institute (*Deutsches Institut für Normung*—DIN), known as DIN-4108, that sets clear rules for calculating Q_P. This takes into account:

- the volume of the building and its floor area
- the area of the building envelope
- a standard ventilation and air leakage factor of 0.7 times the volume of the dwelling(s) per hour
- a solar gain factor based on the building's orientation to the sun
- a standard factor for heat energy gain through indoor activities
- a factor based on the efficiency of the heating system
- the assumption that all rooms in the dwelling(s) are kept at 19 °C throughout the heating season
- the assumption that the average number of heating degree days for Germany applies to the dwelling, wherever it is in Germany.

Standardizing the calculation in this way makes Q_P a useful measure of comparison between buildings or dwellings, and gives project managers a clear target to aim for in their thermal retrofit designs.

However, this measure has come to be used in government and other promotional literature as if it is the actual heating energy consumption of residential buildings, and therefore as a basis for calculating the savings that will be attained through retrofitting. But Q_P is not, actually, the figure that determines householders' energy bills. As we noted in Chap. 2 and will consider in more detail in Chap. 5, actual consumption is, on average, 30% below Q_P, and the percentage gap increases, on average, as Q_P increases. The data considered in Chap. 5 also shows that, for any particular value of Q_P, there is a wide spread of values of actual heating energy consumption. Hence for any particular dwelling the pre- and post-renovation values for Q_P are highly unlikely to give a measure of the likely energy savings (see Chap. 6 for the economic implications of this). Q_P should only be used as a metric to give an overall value to the thermal quality of a dwelling. Although its units are those of heating energy consumption (kWh/m^2a), in the real world it has very little to do with the amount of energy a dwelling actually consumes, or with the fuel bill.

4.4   Practical Problems of Thermal Retrofits in the German Housing Stock

The residential buildings that policymakers want to see thermally renovated were not designed with energy-saving upgrades in mind, and they are a very diverse set of objects. Some aspects of a high proportion of old buildings do lend themselves to easy, cost-effective thermal upgrade measures, but a number of problems are ongoing sticking points. The main ones are: the law of diminishing returns; the wall-roof dilemma; fragile new walls; thermal bridges; and ventilation.

4.4.1   The Law of Diminishing Returns

Using a large database of Swiss home renovation projects, Jakob (2006) showed that the cost of thermal upgrades, per unit of energy saved, rises steeply as the amount of energy saved increases. Galvin (2010a) and Enseling and Hinz (2006) showed the same effect for Germany, though with smaller data samples. There are practical and purely mathematical reasons as to why this occurs.

The mathematical reason has to do with the effect of adding increasing thicknesses of insulation. Suppose, for example, an existing house has $U = 2.0$ W/m^2K for the walls and roof, $U = 5.0$ W/m^2K for windows and external doors, and no serious air leakage problems. Suppose the area of the building envelope is 225 m^2, and the windows and doors make up 25% of this. This gives an H_T value of 2.75 W/m^2K (since $0.75 \times 2.0 + 0.25 \times 5 = 2.75$). Adding a layer of 8 cm of

Styropor will give the wall and roof a U-value of 0.44 W/m^2K, reducing the house's H_T to 1.58 W/m^2K. If the pre-retrofit heating consumption was 280 kWh/m^2a, the thermal upgrade would reduce this to approximately 161 kWh/m^2a (since 280 × 1.58/2.75 = 161), an improvement of 43%.

What will happen if we double the thickness of the wall insulation? This will approximately halve the U-value of the walls and roof, to 0.22 W/m^2K. But the overall transmission loss, HT, will be reduced only to 1.42 W/m^2K, and the heating consumption to 145 kWh/m^2a: an improvement of 10% for a doubling of the wall insulation thickness. Adding a further 8 cm will reduce the walls' and roof's U-value to 0.146 W/m^2K, the building's H_T to 1.36 W/m^2K, and the heating consumption to 138 kWh/m^2a, a further improvement of 5%. Increasing the thickness by yet another 8 cm will produce U = 0.11 W/m^2K, H_T = 1.33 W/m^2K, and heating consumption = 135 kWh/m^2a, a further improvement of 2%. These results are displayed graphically in Fig. 4.5.

One of the reasons the gain in savings flattens out is that the effect of the fixed U-value of the windows and doors comes to dominate, as the wall and roof insulation improves. In a real house the gain in savings would be even flatter, because air leakage, boiler inefficiencies, and solar gain also remain unchanged as insulation thickness increases (hence it is sensible to replace doors, windows, and the boiler when insulating the walls and roof, and to find the optimum combination for maximum saving for minimum cost).

But even if the U-values were reduced for the whole building envelope, the law of diminishing returns would still persist. Each halving of the U-value (which requires a doubling of the insulation thickness) halves the heat loss, so that the thickness of the building envelope increases exponentially while the gain in savings reduces exponentially. This is a mathematical reality that no amount of technical skill can change.

4.4.2 The Wall and Roof Dilemma

Wall insulation is almost always fixed to the outside surface of walls in Germany. There is a limit to how thick this layer may be, before other parts of the house have to be redesigned and rebuilt to accommodate it. Most common is what we call the 'wall-roof dilemma'. Figure 4.6 shows a schematic of a house with a roof over-hang of 10 cm. The EnEV regulations require a 16 cm layer of insulation if the wall only is being repaired or renovated, or a 12 cm layer if the building is being comprehensively thermally renovated (EnEV 2009).

It is not possible to attach 12 cm of external wall insulation to this house, as the insulation would protrude beyond the roof overhang, and there still needs to be room for guttering. The only solution is to extend the roof overhang to at least 20 cm. But there is still a problem, as loft or roof insulation cannot be fitted into the small or non-existent space where the roof meets the wall. This would result in a thermal bridge, with heat escaping freely at the top corner of the house, leading

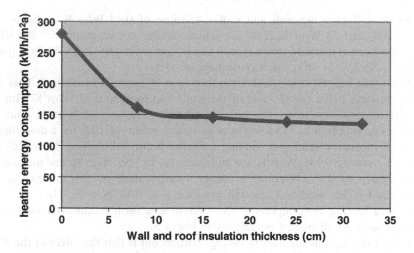

Fig. 4.5 Effect on heating consumption of increasingly thick layers of Styropor® wall and roof insulation being fixed to house with windows and external doors covering 25% of building envelope. Other thermal effects, such as air leakage, solar gain and boiler quality, are ignored

to condensation and mould growth at this corner indoors. The solution here is to raise the roof sufficiently to widen this gap and fill it with insulation. Figure 4.7 shows a schematic of the solution.

Re-modeling the roof adds to the cost of thermal refurbishments, especially if the roof itself was not in need of major maintenance or renewing. Roof overhang lengths vary greatly in Germany, from no overhang at all up to 40 cm. Based on author observations in hundreds of towns, villages, and cities in Germany over the past four years, it seems that only a very small proportion of German homes have roof overhangs on all sides wide enough to accommodate 12 cm of insulation (see Fig. 4.8). As policymakers have steadily increased the required thickness of wall insulation there has come a point for many buildings where roof re-modeling has

Fig. 4.6 Schematic of a house with 10 cm roof overhang

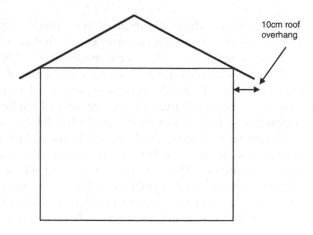

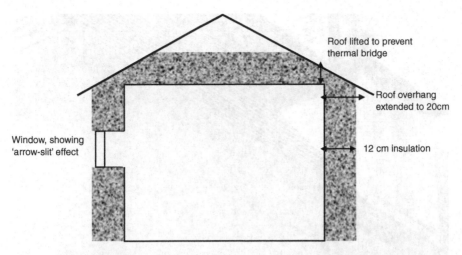

Fig. 4.7 Thermally effective solution, with roof lifted and extended to allow space for continuous sheath of insulation

become necessary. This causes a step-change in the otherwise smooth curve of cost increase with quality of thermal retrofit.

A related problem can occur with basement ceiling insulation. Attaching 10 cm to the basement ceiling (see Table 4.1 in Sect. 4.3.1) reduces the height of the rooms and can reduce their usability. Further, narrow balconies can become unusable when 16 cm of insulation is attached to the wall. In interviews in post-retrofit apartments in Cottbus, Brandenburg, some residents complained that after extra insulation, the balconies were now only useful as storage spaces and homes for pot-plants. External wall insulation can also protrude into driveways and paths, and neighbors' properties if a portion of the external wall is right on the property boundary. From our discussions with the German Energy Agency we learnt that plans to carry out thermal retrofits in this situation can lead to protracted legal disputes.

A further issue if the wall insulation reaches a critical thickness is a kind of 'tunnel' or 'arrow-slit' effect with windows. This is shown schematically in Fig. 4.7 on the left-hand wall. Attempts to achieve very high thermal standards often result in long tunnels for window openings, blocking sunlight, especially where the windows are small. In Munich, for example, one of the municipal buildings was thermally renovated to passive house standard, requiring external wall insulation almost half a metre thick. Its small windows have problems catching light from outdoors.

4.4.3 Fragile External Wall Insulation

A further challenge many German homeowners are facing with external wall insulation is its fragility. Styropor, Neopor, other forms of polystyrene, and nonpetroleum products such as rock-wool have very low tensile and compressive strength.

Fig. 4.8 Most German houses do not have overhangs on all sides wide enough to accommodate 12 cm of extra insulation as required by the EnEV

They are glued and screwed to external walls in blocks of around 30 cm × 15 cm × 16 cm, then covered with a fiberglass coating to provide a hard surface, and finally plastered over (see Figs. 4.2 and 4.3 in Sect. 4.2.1). This configuration has very low impact resistance, and can be damaged by children, bicycles leaned against a wall, or woodpeckers eating insects clinging within the stucco folds of the plaster (Handwerk 2009). The plaster can crack with time and weather. In the calculations of economic viability (see Chaps. 2 and 6) external wall insulation is optimistically deemed to last 25 years. Discussions with leading personnel at Munich City Council's Building Advice Center (www.muenchen.de/bauzentrum) revealed a scepticism, among experienced builders, as to the longevity of this form of insulation: it could last anything from 3–25 years.

4.4.4 Ventilation and Mould Risks

Many occupants complain that mould begins to form in homes that have been thermally renovated. There is a large corpus of literature on mould formation in homes: its possible causes (e.g. Engman et al. 2007; Howden-Chapman et al. 2005) its implications for health (e.g. Singh 2001), and ways of alleviating it (e.g. Galvin 2010b). There is also much folklore in Germany on mould, in particular the view that it forms indoors when a wall 'cannot breathe' (e.g. SZ Süddeutsche Zeitung 2012). The assumption is that air and moisture should be able to pass freely through a wall, and when this cannot happen, moisture builds up on the indoor surface and leads to mould growth. We find no support for this view in scientific research, and there are more obvious reasons as to why moisture comes to settle on indoor surfaces of walls.

Human activities cause an increase in the moisture content, and therefore the relative humidity, of indoor air during the day. As this air cools down at night its relative humidity rises. If any pockets of air cool below the dew point, water will condense out of these pockets. Usually the coldest air is alongside the coolest surfaces, which are usually window panes and the frames and structure around windows, the lower reaches of the wall near the floor and three-way corners which are often thermal bridges by nature of their geometry (Galvin 2010b). Moisture, cool temperatures, and organic material such as wallpaper or clothing provide ideal conditions for mould growth.

There are four different ways this problem can be solved: keep the air warm at night; repair the thermal bridges; dehumidify the air at night; or ventilate sensibly. Most households seem to under-ventilate, which can lead to mould formation, or over-ventilate, which markedly increases heating fuel consumption (Galvin 2013).

4.5 Conclusions and Implications

Domestic space and water heating accounts for approximately 14.6% of Germany's energy consumption. Replacement of older homes with energy-efficient new builds can bring about some reductions in this consumption, but, as we saw in Chap. 2, this is limited by the fact that nearly twice as many new homes are being built as the number abandoned or demolished. There is therefore a focus among policymakers on existing homes as a source of heating energy reduction.

In this chapter we have outlined the technical issues involved in thermally renovating German homes, relating these to basic principles of physics, and to the actual character and peculiarities of these homes.

The physics shows that, while transmission losses can be precisely predicted from the properties of insulation materials, figures typically derived for quantities of heating energy consumed are highly theoretical, based on a series of unknowns, and do not relate well to the actual thermal performance of dwellings either before or after a retrofit. This has led to estimates of current energy consumption, and of the savings potential of retrofits, that do not reflect what actually happens in practice. Policymakers need to review the way these figures are used in their estimation and promotion of the benefits of thermal renovation.

Further, attempts to push thermal renovation to higher and higher standards run up against basic problems of physics and mechanics. There is a law of diminishing returns inherent in the way insulation functions on houses as its thickness becomes greater, so that the advantages fall away rapidly when U-values go below about 0.4, representing a thickness of 8–10 cm of typical insulation material. The mechanics and geometry of attaching thick insulation to existing wall-roof configurations, to basement ceilings, to walls alongside balconies, and around windows make for further limitations in the thermal standards that can be achieved economically. Additional energy-efficiency problems are inherent in homes that have many corners and recesses, or a poor orientation to the sun. Further, eliminating energy wastage from air leaks, by making homes air-tight, leads to the need for adequate ventilation to avoid mould problems, and German householders are proving markedly unskilled in ventilating energy-efficiently.

Over the past decade there has been a dominant discourse among German policymakers and their expert advisors that it is technically and economically possible to reduce heating consumption deeply through thermal renovation of existing homes. This confidence now appears to be waning, as reports come in of practical difficulties in meeting the stringent standards of EnEV 2009, and even the less stringent standards of EnEV 2002. Policymakers need to become intimately conversant with the detailed physical and mechanical features of these problems, so that shifts in policy are based on hard facts rather than merely a response to strong lobby groups.

References

BMVBS Bundesministerium für Verkehr, Bau und Stadtentwicklung (2011) CO$_2$-Gebäudesanierung - Energieeffizient Bauen und Sanieren: Die Fakten. http://www.bmvbs.de/SharedDocs/DE/Artikel/B/co2-gebaeudesanierung-energieeffizient-bauen-und-sanieren-die-fakten.html. Accessed 6 Oct 2012

BMWi (Bundesministerium für Wirtschaft und Technologie) (2010) Energie in Deutschland: Trends und Hintergründe zur Energieversorgung. Aktualisierte Ausgabe Aug 2010. BMWi, Berlin

BMWi (Bundesministerium für Wirtschaft und Technologie) (2011) Energiedaten: ausgewählte Grafiken. Stand: 15.08.2011. BMWi, Berlin

DENA (Deutsche Energie-Agentur) (2010) dena-Sanierungsstudie. Teil 1: Wirtschaftlichkeit energetischer Modernisierung im Mietwohnungsbestand: Begleitforschung zum dena-Projekt „Niedrigenergiehaus im Bestand". DENA, Berlin

EnEV (Energieeinsparverordnung) (2009) EnEV 2009—Energieeinsparverordnung für Gebäude. http://www.enev-online.org/enev_2009_volltext/index.htm. Accessed 26 Sept 2012

Engman LH, Bornehag CG, Sundell J (2007) 'How valid are parents' questionnaire responses regarding building characteristics, mouldy odour, and signs of moisture problems in Swedish homes? Scandinavia J Public Health 35(2):125–132

Enseling A, Hinz E (2006) Energetische Gebäudesanierung und Wirtschaftlichkeit—EineUntersuchung am Beispiel des 'Brunckviertels' in Ludwigshafen. Institut Wohnen und Umwelt, Darmstadt

Galvin R (2010a) Thermal upgrades of existing homes in Germany: the building code, subsidies, and economic efficiency. Energy Build 42(6):834–844

Galvin R (2010b) Solving mould and condensation problems: a dehumidifier trial in a suburban house in Britain. Energy Build 42(11):2118–2123

Galvin R (2013) Impediments to energy-efficient ventilation in German dwellings: a case study in Aachen. Energy Build 56:32–40. doi:10.1016/j.enbuild.2012.10.020

Handwerk (2009) Besitzer älterer Häuser können bei der Sanierung künftig ein Drittel der Baukosten einsparen. Handwerk Magazin: 01.09.2009. http://www.handwerk-magazin.de/data/news/News-Energiesparpraemie-fuer-Wohnhaeuser_4054484.html. Accessed 5 Sept 2009

Howden-Chapman P, Saville-Smith K, Crane J, Wilson N (2005) Risk factors for mold in housing: a national survey. Indoor Air 15(6):469–476

IEA (International Energy Agency) (2008) Promoting energy efficiency investments: case studies in the residential sector. International Energy Agency, Paris

Jakob M (2006) Marginal costs and co-benefits of energy efficiency investments: the case of the swiss residential sector. Energy Policy 34:172–187

McKinsey & Company (2009) Pathways to a Low-Carbon Economy: Version 2 of Global Greenhouse Gas Abatement Cost Curve, McKinsey & Company. https://solutions.mckinsey.com/ClimateDesk/default.aspx. Accessed 19 Dec 2012

Singh J (2001) Review: occupational exposure to moulds in buildings. Indoor Built Environ 10(3–4):172–178

SZ (Süddeutsche Zeitung) (2012) Hausdämmung. Süddeutsche Zeitung Nr. 214 15-16 Sept, 2012: V2: 4–5

Chapter 5
The Prebound Effect: Discrepancies Between Measured and Calculated Consumption

Abstract Estimates of heating fuel saving potential in German homes are generally based on a calculated consumption figure. The methodology for working this out is set down by the German Institute of Standards. But how close is this figure to dwellings' actual heating energy consumption, and how does this affect the real energy savings potential through thermal retrofits? This chapter examines existing data on 3,400 representative German homes, their calculated energy rating plotted against their actual measured consumption. The results indicate that occupants consume, on average, 30% less heating energy than the calculated rating. This phenomenon, which we call the 'prebound' effect, increases with the calculated rating. The opposite phenomenon, the rebound effect, tends to occur for low-energy dwellings, where occupants consume more than the rating. A similar phenomenon has been recognized in recent Dutch, Belgian, French, and UK studies, suggesting policy implications in two directions. First, using a dwelling's energy rating to predict fuel and CO_2 savings through retrofits tends to overestimate savings, underestimate payback time, and possibly discourage cost-effective, incremental improvements. Second, the potential fuel and CO_2 savings through nontechnical measures like occupant behavior may well be far larger than is generally assumed in policies, so policymakers need a better understanding of what drives or inhibits occupants' decisions.

Keywords Heating energy consumption · Energy rating · Occupant behavior · Thermal retrofit · Prebound effect

5.1 Introduction

In Chaps. 2 and 4, we saw that Germany uses two different metrics for the thermal performance of buildings: heat transmission losses (H_T); and the theoretical quantity of primary energy consumed per square metre of indoor 'useful' area per year (Q_P). The first, as we saw, gives a measure of the heat losses, by conduction,

R. Galvin and M. Sunikka-Blank, *A Critical Appraisal of Germany's*   67
Thermal Retrofit Policy, Green Energy and Technology,
DOI: 10.1007/978-1-4471-5367-2_5, © Springer-Verlag London 2013

through the building envelope. It is robustly objective, based purely on physical laws and the properties of insulating materials, but is of limited value for evaluating overall building performance, because conduction of heat through the building envelope is only one of a number of factors determining how much heating energy a building consumes. The second metric gives a fuller overall picture, but as we saw in Sect. 4.3, much of it is based on assumptions and standardizations that might not be true for real dwellings with real people in them.

We argued that Q_P should therefore only be used as a proxy measure for comparing the theoretical performance of buildings: it cannot predict how much energy a home actually consumes, nor its annual heating bill.

One area of confusion in this regard is energy performance certificates (EPCs). There are two types of EPCs in Germany: the *Energiebedarfsausweis* (energy demand certificate) and the *Energieverbrauchsausweis* (energy consumption certificate). The former uses Q_P as the dwelling's energy rating. The latter uses an actual, measured energy consumption figure, which must be averaged from the last three years' energy bills and adjusted to give primary energy consumption. The certificates can be distinguished by the word fragment '-*bedarf*' (demand) or '-*verbrauch*' (consumption) in their titles. The -*bedarf* version, based on Q_P, is compulsory for new builds and for buildings of 1–4 dwellings built before 1 November 1977 that have not been thermally renovated to the standard of a post-1977 new build. All other buildings may have either a -*bedarf* or -*verbrauch* version. The -*verbrauch* version is rare, because occupants do not always have 3 years of energy bills to hand. If there is any significant discrepancy between Q_P and actual, measured performance, EPCs can be as much a source of confusion as of information.

But how risky is it to use Q_P as an estimate of consumption? How far removed is Q_P from actual, measured energy consumption in German homes? Walberg et al. (2011), who investigated the actual heating consumption and thermal retrofit potential of tens of thousands of homes from multiple data sources in Germany, concluded:

> For a realistic assessment of the thermal condition of the built environment only the analysis of actual, measured energy consumption can be used... Theoretically calculated energy ratings give us an unrealistic picture of the energy savings potential that can be achieved through thermal renovation. (Walberg et al. 2011, p. 115; authors' translation)

Greller et al. (2010), using measured data from over 110,000 rented apartments throughout Germany, found that older buildings are consuming significantly less energy than their 'reputation' (i.e. their calculated Q_P), while new buildings are consuming significantly more (ibid: 5). Schröder et al. (2011), using metered data from over 1,000,000 dwellings, concluded that using calculated energy ratings leads to overestimation of the energy saving potential of residential buildings (ibid: 253). CO$_2$Online (http://www.co2online.de), which collects and analyzes data on energy performance of dwellings from over 1,000,000 German households, also expresses concern that Q_P is generally significantly higher than these dwellings' actual consumption[1] (but see note in Sect. 5.2 on this).

As we saw in Chap. 4, the official methodology for calculating Q_P, set by the German Institute of Standards, makes standard assumptions about household heating behavior: that occupants heat all rooms constantly to 19 °C, and ventilate at the rate of 0.7 times the volume of the dwelling per hour. Making such assumptions is a necessary step in producing a figure that indicates the quality of the building itself: all buildings must be based on the same occupant behavior to give a fair comparison. But in the real world, occupants vary enormously, and this limits the usefulness of Q_P as a measure of actual performance. A number of studies over the last 12 years, from a range of countries, have shown that differences in occupant heating behavior are at least as much a determinant of heating consumption as is the thermal quality of the building (Stern 2000; Guerra Santin et al. 2009; Gram-Hanssen 2011). Haas and Biermayr (2000) showed that, due the behavior factor, heating energy saving achieved through retrofit measures can be remarkably lower than calculated.

In this chapter, we investigate the relationship between German dwellings' calculated thermal performance (Q_P), and their actual, measured performance (adjusted to indicate primary energy consumption)—which, for the sake of brevity, we will frequently call 'Q_M'. We approach this investigation with three questions in mind:

1. In what ways do Q_P and Q_M vary in relation to each other?
2. What can be learnt from this about the heating energy savings potential in German dwellings?
3. What does this suggest for the policy aim of achieving large savings in heating energy and CO_2 emissions from thermal retrofits?

In Sect. 5.2, we analyze a number of existing studies that compare Q_P with Q_M for large datasets of residential buildings. In Sect. 5.3, we explore possible reasons for discrepancies between these two metrics. Section 5.4 introduces the notion of the 'prebound effect' to help conceptualize the way Q_P and Q_M vary in relation to each other, contrasting this with the well-known concept of the 'rebound effect'. Section 5.5 compares our findings for Germany with comparable results in the Netherlands, the UK, Belgium and France, and Sect. 5.6 offers conclusions and recommendations for policymakers and those undertaking building renovation.

---

[1] Personal communication with Dr Johannes Hengstenberg, Director of CO_2Online, October 10, 2012.

5.2 The Actual Measured Energy Consumption in German Dwellings

Over the last decade in Germany, a number of studies have compared the calculated heating energy consumption of dwellings (Q_P), with the actual, measured consumption (Q_M) (Kaßner et al. 2010; Knissel et al. 2006; Knissel and Loga 2006; Greller et al. 2010; Loga et al. 2011; Erhorn 2007; Jagnow and Wolf 2008). These studies reveal and quantify discrepancies between Q_P and Q_M, though they were mostly undertaken for purposes different from our aim in this chapter: some investigate whether we can use measured energy consumption as an inexpensive way to estimate Q_P (e.g. Knissel et al. 2006; Knissel and Loga 2006); some ask how various building typologies perform in practice (Greller et al. 2010; Loga et al. 2011); or how energy advisors can better inform consumers of the energy-saving potential of their properties (Erhorn 2007). In addition, several recent empirical studies have quantified average actual energy consumption (Q_M) for various classes of residential building (Schloman et al. 2004; Schröder et al. 2010; Walberg et al. 2011) and for the residential building stock as a whole (Schröder et al. 2011), and we can compare their results with average values of Q_P for such buildings.

By gathering the datasets in these studies together and exploring commonalities and differences among them, we can get a picture how Q_P and Q_M compare with each other in a range of types and thermal qualities of dwellings. The studies cover direct observations of actual energy consumption (Q_M) in over 1,000,000 dwellings, and 3,400 dwellings with precise information on both Q_P and Q_M, where direct, dwelling-by-dwelling comparisons are made. Measured data was generally collected from meters during a one to four-year period, and cooling is not included. In all cases, measured data, which gives final energy consumption, has been adjusted to indicate primary energy consumption.

Measures were taken in our analysis to ensure the quality of the sources and correct interpretation of results. Only work by established research institutes and peer-reviewed reports were used as data sources, and only where there was transparency in data collection method and the calculations used in data processing. For example, we have not used data from CO_2Online, even though this offers direct comparisons of Q_P and Q_M in tens of thousands of dwellings, as there could be a self-selection bias in this data. It is not randomly selected but collected from volunteer households who submit it via an Internet portal. Those who volunteer are likely to be more energy conscious than average, and may therefore be more likely to be consuming less energy than their dwellings' Q_P rating. Box 5.1 gives an overview of some of the issues we need to be aware of when using secondary data in this way.

Box 5.1 Alternative Metrics for Heating Energy Consumption

There are three sets of alternatives that need to be taken into account when stating the calculated heating energy consumption, per square metre of floor area, of German residential buildings:

1. Is water heating included?
2. Are we referring to 'primary' or 'final' energy consumption?
3. Is the floor area the total for the building, or just the area inside the dwellings?

1. Water heating is generally considered to add around 12.5 kWh/m²a to a households' heating energy consumption, and it is included in all our discussion unless otherwise stated.

2. The subscript $_P$ in the term Q_P stands for calculated *primary* energy consumption, i.e. the total energy required, from nonrenewable sources, for the building's space and water heating including fuel transport (e.g., for oil) and production and transmission inefficiencies (e.g. losses in electricity generators and grids). An alternative metric, Q_E, is the calculated *final* energy consumption, which does not include these transport and production losses. Both metrics take account of the natural gains through sun and human indoor activities, and standardized factors for heating and ventilating behavior. EnEV 2002 and 2009 use Q_P, making their requirements around 10% stricter than they would be if based on Q_E for dwellings using oil or natural gas, but 2.6 times as strict for dwellings heating with electricity. Wood fuel is given a nominal value of $Q_P/Q_E = 0.2$.

3. In reading German reports of research on the heating energy consumption of residential buildings, we have to be careful to check whether these refer to consumption per square metre of *living* area *(Wohnfläche)* or *useable* area *(Nutzfläche)*. Living area refers only to the floor area inside the apartments. Useable area includes the public areas of the building (stairs, corridors, drying rooms, basement etc.). EnEV 2009 includes these areas in the methods it prescribes for calculating Q_P. This puts the paper value of Q_P lower than it would be if it only included living area (since we are dividing kilowatt-hours by a larger number of square metres). Hence if the public areas of a building make up 10% of its floor area, a dwelling with a Q_P rating of 70 kWh/m²a will actually be consuming 77 kWh/m²a.

Different German studies use different combinations of these three factors in their analyses. When comparing calculated with actual heating consumption within a given dataset this is not usually an insurmountable problem, provided each study is internally consistent.

In the cases where raw data were not available, our analysis is based on studies' statistical summaries and graphical data presentation. Due to our long involvement with German policy and practice in thermal renovation of existing homes, we were able to check the reasonableness of results against a wide range of background knowledge and discussions with practitioners (see, e.g. Galvin 2012).

Four interesting features turned out to be common to these datasets.

First, for any given energy rating (Q_P) the data show a very large spread of quantities of energy consumed for heating (Q_M). Typical are ranges of over 600%, i.e. one home consumes six times as much energy for heating as another of the same thermal rating (Erhorn 2007; Knissel and Loga 2006; Loga et al. 2011). This phenomenon is not specific to Germany. It is also evident, for example, in Switzerland (Jakob 2007), France (Cayre et al. 2011; Cayla et al. 2010), Austria (Roth and Engelman 2010), the Netherlands (Tighelaar and Menkveld 2011), and Denmark (Erhorn 2007).

Second, in all the datasets there is a significant gap between the average value of Q_P and the average value of Q_M. The average value for Q_P is around 225 kWh/m^2a, with a range from 15 kWh/m^2a to over 400 kWh/m^2a. By contrast, the average value for Q_M is around 150 kWh/m^2a, with a range from 10 kWh/m^2a to not much over 300 kWh/m^2a. For example, Schröder et al. (2011) estimate the average for Q_M at 149 kWh/m^2a, based on metered data from over 1,000,000 dwellings throughout Germany, and Walberg et al. (2011), using similarly large datasets, estimate it at 152 kWh/m^2a.

Hence, the datasets indicate that Q_M is, on average, 30% lower than Q_P. This can be expressed in the alternate form: Q_P is, on average, 40% higher than Q_M.

Third, the datasets suggest a trend in energy consumption in relation to the magnitude of Q_P. In general, the higher the value of Q_P, the larger the percentage gap between Q_P and average Q_M. For example, the average measured consumption (Q_M) of dwellings with a calculated consumption (Q_P) of 300 kWh/m^2a is around 40% below Q_P, while dwellings with an average Q_P of 150 kWh/m^2a show an actual energy consumption (Q_M) around 17% below Q_P.

These points are illustrated in Fig. 5.1, which shows scatterplots of Q_M (vertical axis) against Q_P (horizontal axis) for detached houses (left) and multi-dwelling buildings (right), from data collected by the German Energy Agency and analyzed by Erhorn (2007). In each graph the continuous line is the regression line, while the dotted line is y = x, representing all the points where Q_M, the actual consumption, would be identical to the calculated Q_P value. The wide vertical scattering of the points at any particular x-value reflects the large variation in energy use regardless of the physical features of the building, while the general shape of the regression line indicates approximately how the average value of Q_M varies as Q_P varies.

Plots such as these suggest that many households are consuming significantly less than their calculated rating. The left hand graph in Fig. 5.1 shows some detached houses consuming only around 100–130 kWh/m^2a even though their Q_P is up to 400 kWh/m^2a.

Other studies suggest similar trends in German homes (Kaßner et al. 2010; Knissel and Loga 2006; Knissel et al. 2006; Jagnow and Wolf 2008; Loga et al. 2011). For dwellings with Q_P above 100 kWh/m^2a the effect is similar in all these studies. Further, the x-coefficient of the regression line ranges from 0.2 to 0.5,

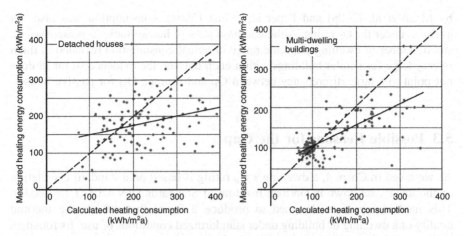

Fig. 5.1 Scatterplots of measured energy consumption (*vertical axes*) against calculated energy consumption (*horizontal axes*) for detached houses (*left*) and multi-dwelling buildings (*right*). *Source* Data from Erhorn (2007). *Solid line* is line-of-best-fit; *dotted line* is y – x, where measured consumption = calculated consumption

indicating that for each 1% increase in the thermal leakiness of these homes there is only a 0.2 to 0.5% increase in heating energy consumption.

Fourth, at the other end of the scale, low-energy dwellings generally indicate the opposite tendency. The right side graph of Fig. 5.1 shows the majority of points at the low energy end falling above the line y = x, indicating that in many low-energy dwellings the actual consumption exceeds the calculated rating. Other datasets support this tendency. Loga et al. (2011) show average measured consumption rising above Q_P for dwellings with Q_P lower than 50 kWh/m²a. This is more pronounced (around 65%) for dwellings with Q_P under 75 kWh/m²a in datasets from Kaßner et al. (2010). Other analyses show a similar although lesser tendency (Jagnow and Wolf 2008; Knissel et al. 2006). In a study of neighborhoods of what Thomsen et al. (2005) call 'advanced solar low energy buildings', actual measured heating energy consumption was measured to be twice as high as Q_P. Greller et al. (2010) investigated the actual heating energy consumption in German buildings according to year of build, and found that, as thermal standards have tightened progressively in recent years, a larger and larger proportion of new builds has failed to achieve the required thermal standard: Q_M is rising more and more above Q_P. An exception is a study of a small, homogeneous sample by Enseling and Hinz (2006), where the average post-retrofit consumption in low-energy retrofits falls within their Q_P rating.

With passive houses, however, the tendency seems less consistent. Berndgen-Kaiser et al. (2007) surveyed the performance of 700 passive and 370 low-energy houses in North Rhine-Westphalia. The measured heating energy consumption for 57% of the low-energy houses was above the Q_P rating, but this was so for only and 41% of the passive houses. In studies of small passive house estates conducted by Maaß et al. (2008) and Peper and Feist (2008), consumption was also, on average, under the Q_P rating. Although these passive house studies represent only a small number of dwellings, and the range of actual consumption revealed in them is very large for similar buildings within each dataset, the evidence so far in does not point to a large discrepancy between Q_P and average Q_M for passive houses.

5.3 Possible Reasons for the Gap

As we noted in Chap. 4, a dwelling's Q_P rating is based on a standard calculation methodology, set down in German Institute of Standards DIN 4108 (DIN 2003a). This methodology was designed to produce a figure that reflects the thermal quality of a dwelling or building under standardized conditions of use. Its founders never expected any particular dwelling with particular occupants to perform precisely, or even approximately, according to the Q_P rating. A large range of actual consumption values for any particular Q_P rating was always expected, though it was also expected that the average of these would be close to Q_P. Another German Institute of Standards technical advice publication, DIN EN 832:1998 (DIN 2003b), points out that the calculation of Q_P can vary by 50–150% depending on the assumptions made about user behavior, including ventilation behavior (quoted in Beecken and Schulze 2011: 340).

The surprises, however, are that average actual consumption is well below Q_P; that the percentage gap increases as Q_P rises; and that this does not seem to apply to passive houses.

Part of the reason for the large gap could be that the thermal properties of many older homes have been wrongly assessed when their Q_P rating has been calculated. However, it is unlikely that the inaccuracies could lead to an average 30% error, for two main reasons. First, it is not difficult to see what a building is made of, how thick its walls and roof are, how draughty it is, what thermal quality its windows have, and how efficient its heating system is. Even if there are inaccuracies in measurement of wall thicknesses and boiler efficiencies and so on, it is highly unlikely that they would be consistently on the high side and of such magnitude.

Second, we would still need to explain why the gap increases as Q_P rises, but becomes negative for low values of Q_P.

Another reason for the large gap could simply be that the user assumptions in DIN 4108 are unrealistic: that homes do not really ventilate at 0.7 full exchanges of air per hour; and people do not keep all the rooms constantly at 19 °C or above. This might help explain why the average gap is so large, at 30%, but again it would not explain why it rises as Q_P rises, yet falls to 0% at around 100 kWh/m^{2}a and goes negative below that Q_P rating.

One factor could be the behavior of the occupants. It appears that, the higher the Q_P rating (i.e. the thermally leakier the building) the more economical the occupants' heating behavior becomes. We may hypothesize that householders heat fewer rooms, or to a lower temperature, or for fewer hours per day, or various combinations of these strategies, and possibly also ventilate more sparingly, when they are aware that their consumption would otherwise be excessive. Simons (2012) draws the same conclusion in examining similar data on 7,510 detached and semi-detached houses throughout Germany. Here, the average calculated consumption (Q_P) for nonretrofitted homes was 400 kWh/m²a while the average measured consumption was 167 kWh/m²a.

This could suggest there is fuel poverty in many German homes, as people respond to the cost of heating a thermally leaky home by heating it less than is good for their health and wellbeing. On the other hand, many homes in Germany are only partially occupied, as the population and the number of persons per household are falling in Germany while the total number of dwellings is increasing (BMVBS 2011). Further, in many homes all the occupants are out at work or education during the day, so that heating and ventilation are not required. These factors offer many opportunities for fuel saving. The graphs of actual consumption against Q_P suggest that people in thermally leaky homes are more likely to take advantage of them.

This would also help explain why the average actual, measured rating goes above Q_P for Q_P ratings below 100 kWh/m²a: when householders are aware that their consumption is relatively low, they heat more liberally. But why do passive houses seem exempt from this trend? This could be due to the type of heating system these houses have: a small heat pump embedded in the ventilation system. This contrasts with the traditional heating system of other low-energy houses, where a boiler serves radiators in all the rooms. It may be easier to overheat with a traditional, high-capacity boiler feeding radiators throughout the house, than with a small heat pump. Alternatively, passive house samples might be weighted by the possibility that people who choose to buy a passive house are more likely to be energy-saving conscious. One study found that people who planned to build or buy a passive house consumed, on average, significantly less heating energy than people who were persuaded by an architect to build one, or who were placed in a passive house by a housing provider (ILS 2010: 34ff).

5.4 The Prebound Effect in Household Energy Consumption

The analysis above suggests that actual measured household heating energy consumption in German homes is on average 30% lower than the dwellings' calculated Q_P rating, and that, in general, the worse a home is thermally, the more economically the occupants tend to behave with respect to their space heating. We call this the 'prebound' effect, as it is the opposite of the 'rebound' effect (Sunikka-Blank and Galvin 2012).

The 'rebound' effect occurs when a proportion of the energy saved through energy-efficiency upgrades is consumed by additional energy usage. It was originally identified by Jevons (1865), and brought into recent discussion by Brookes (1979) and Khazzoom (1980). It has been identified across a wide range of energy consumption sectors (Barker et al. 2007; Holm and Englund 2009; Sorrell and Dimitropoulos 2008), including home heating (e.g. Haas and Biermayr 2000). After a thermal retrofit, many occupants use more energy than the calculated rating (their dwellings' new Q_P), for example by heating more rooms, keeping higher temperatures, heating for longer periods, and ventilating more generously. The data we examined above suggest that this is happening for most homes with Q_P below 100 kWh/m^{2}a, and more intensely as Q_P goes lower.

By contrast, the 'prebound' effect is the tendency to consume less energy than the calculated rating. The datasets examined above indicate that this occurs in thermally less efficient homes, prior to or in the absence of thermal upgrade measures, and roughly in proportion to the thermal leakiness of the building. Because most German homes have a Q_P rating well above 100 kWh/m^{2}a, the prebound effect is far more dominant in the housing sector than the rebound effect. Many studies invoke the rebound effect to explain why the fuel savings achieved through thermal upgrades are often much smaller than anticipated. A more salient reason, however, might be the prebound effect: homes cannot save energy they were not previously consuming.

Although the datasets we examined were from independent sources, they show a remarkable consistency in the general way the magnitude of the prebound effect varies in relation to Q_P. For example, Loga et al. (2011) offer a general curve to display the relationship between Q_P and measured heating energy consumption. In their dataset, the average actual space heating consumption of dwellings with a Q_P of 500 kWh/m^{2}a is around 215 kWh/m^{2}a, while that for dwellings with a Q_P of 200 kWh/m^{2}a is around 145 kWh/m^{2}a—though there are obvious variations across types, sizes, and ages of dwelling. Building on Loga et al.'s (2011) curve fitting equation we can develop a model to describe the prebound effect, namely:

$$P = 100[1.2 - 1.3/(1 + Q_P/500)],$$

expressed as a percentage.[2]

This is displayed in Fig. 5.2. We note that, in this model, the prebound effect becomes zero where $Q_P = 50$ kWh/m^{2}a, and for Q_P below this it is negative, i.e. the rebound effect dominates. The zero point varies between datasets. In those from Erhorn (2007), it becomes negative for detached houses when Q_P goes below 150 kWh/m^{2}a, and for multi-apartment buildings where Q_P is below 100 kWh/m^{2}a. However, the general shape of the curve is the same in all cases.

---

[2] The numbers 1.2, 1.3 and 500 in the equation are modeling coefficients, arrived at by trial and error to produce a smooth curve that fits best fit with the data. The model used here is a reciprocal function, but there is no necessary reason why a logarithmic or power function (as in Fig. 5.3) could not have been used, in which case different modeling coefficients would have been arrived at.

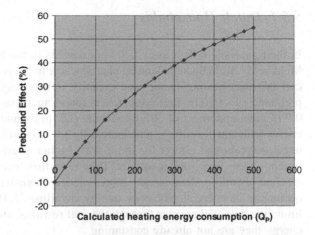

Fig. 5.2 Graphical model of prebound effect as function of calculated heating energy consumption in German homes, from dataset in Loga et al. (2011). $P = 100 \, [1.2 - 1.3/(1 + Q_P/500)]$, expressed as a percentage

This type of curve can offer a quick, approximate reference for the average actual consumption (Q_M) to be expected in dwellings with any particular value of Q_P, though again we must note that it plots average values for Q_M for each value of Q_P, so that the Q_M for any particular home might be significantly above or below this.

Using such curves can also throw light on the savings potential to be expected through thermally upgrading large numbers of homes of any particular Q_P rating. For example, suppose we renovate a number of homes with a Q_P of around 225 kWh/m²a (the approximate national average) to a standard that aims to produce a new Q_P of around 100 kWh/m²a (the current legal maximum). This would represent a fuel consumption saving of 56%. However, the prebound effect suggests that the actual preretrofit consumption of these homes (Q_M) averages 30% below 225 kWh/m²a, i.e. 158 kWh/m²a. Hence, provided the post-retrofit actual consumption does reach 100 kWh/m²a (i.e. there is no rebound effect), the actual saving is only 58 kWh/m²a. This is a 37% saving in actual fuel consumption, significantly less than the 55% expected. Gaps this large could have serious implications for fuel and GHG reduction policies, and for the budgeting of private homeowners and commercial and state housing providers.

5.5 Evidence from Other European Countries

When we identified the prebound effect in Germany, we examined wider international research to see if it was evident in other European countries. We found a similar phenomenon in studies of Dutch, French, Belgian and UK homes, and compared our findings to those of researchers working in these countries in a workshop in Amsterdam in January 2012.

5.5.1 Dutch Households

In their analysis of data from 4,700 households in the Netherlands, Tighelaar and Menkveld (2011) find an identical phenomenon to the prebound effect, though they call this the 'heating factor' and quantify it inversely (heating factor = 1.0 − prebound effect/100). These researchers found an average heating factor of around 0.7 (a prebound effect of 30%), reducing (i.e. prebound effect increasing) for less energy efficient dwellings, and increasing to 1.0 (prebound = 0%) or higher (a negative prebound effect) for those with higher energy efficiency. Hence, their findings were very similar to ours. They note that 'occupants in an energetically efficient dwelling demonstrate more energy intensive behaviour compared to occupants in energetically poor quality dwellings'. They suggest this severely limits the potential savings through thermal retrofits, since occupants cannot save energy they are not already consuming.

5.5.2 UK Households

In Kelly's (2011) study of the UK housing stock, a correlation was found between dwellings' energy efficiency and their energy demand. Dwellings' energy efficiency is expressed in the UK as a 'Standard Assessment Procedure' (SAP) rating on a scale of 1–100, where 100 is the most energy efficient and 1 is the least (the opposite direction of the Q_P scale). Kelly analyzes data from the English House Condition Survey (EHCS) of 2,531 dwellings, using a structural equation model that enables cross-correlations to be examined for a range of factors likely to be associated with each other in relation to heating demand. He uses the notion of 'propensity to consume more (or less) energy' for factors that are not dependent on the dwellings' physical thermal characteristics (indoor temperature, floor area, number of occupants, and income level) all of which, he finds, are positively correlated with energy consumption. Homes with a high SAP have a 'propensity to consume more energy', while the opposite is the case for homes with a low SAP. In other words, the worse the energy rating, the lower the energy consumption in relation to the rating—as with German households (see Sect. 5.4). Kelly suggests that, on the one hand, the costs of further thermal improvements in homes with a high SAP rating will be high due to the law of diminishing returns (compare Jakob 2006); while on the other hand, retrofitting homes with a low SAP rating may lead to increases in average internal temperature rather than decreases in energy consumption.

5.5.3 Belgian Households

Hens et al. (2010) analyzed a dataset of building characteristics and measured heating fuel consumption of 964 Belgian dwellings with known heat transmission loss figures. As an independent variable they used an interesting metric for dwellings' energy rating, which is closer to the German H_T (heat losses through the building envelope, in W/m^2K—see Chap. 4) than Q_P (theoretical heat energy consumed, in kWh/m^2a). This is 'specific transmission losses per m^3 of protected volume' (STV), expressed in W/m^3K. It is the average U-value of the building envelope, divided by the building's volume and multiplied by the building envelope's area. This takes account of the higher thermal quality afforded by having a lower ratio of building envelope area to volume of building (see discussion in Chap. 4, Sect. 4.2).

For their dependent variable, these researchers used 'heating energy consumption per unit volume' of dwelling, rather than per m^2 of indoor area. This corrects for variations in energy consumption due to different ceiling heights. They plotted this against the STV for their 964 exemplars.

We note the similarity between this and the plots of heating energy consumed against Q_P for German dwellings (see Fig. 5.2). Hens et al. (2010) use curve fitting with this plot to derive an equation for what they call the rebound effect, though more correctly it is what we call the 'prebound' effect (see Sect. 4.4), as it maps the percentage by which the actual energy consumption falls below the calculated value:

$$P = 100 \left[1.355(U/C)^{0.16} - 1 \right]$$

where U = transmission loss (W/m^2K), and C = compactness of building: i.e. volume/area of building envelope.[3]

This curve is displayed in Fig. 5.3. It has a similar form to the one we derived from Loga et al. (2010) (Fig. 5.2). The higher the specific transmission losses per m^3, the larger the proportionate gap between measured energy and this variable. The equation also enables modeling to extend into the low transmission loss (= high energy efficiency) area where the prebound effect becomes negative, i.e. the rebound effect becomes dominant.

5.5.4 French Households

Cayre et al. (2011) relate measured space heating energy consumption to the 'Energy Performance Certificate' (EPC) value for space heating in French households. Instead of using figures for kWh/m^2a, the EPC is given in MWh per dwelling per year (MWh/dw.y). This makes a comparison with German data somewhat difficult, as it tends to conflate the effects of size of dwelling, with thermal leakiness of dwelling: a large, thermally efficient dwelling may have the

---

[3] Again the figures 1.355 and 0.16 are modeling coefficients, arrived at by trial and error, to produce a curve that best fits the data.

Fig. 5.3 Prebound effect as a function of transmission losses per unit volume in Belgian dataset. $P = 100$ $[1.355\ (U/C)^{0.16} - 1]$, where U = transmission loss (W/m^2K), and C = compactness of building: i.e. volume/area of building envelope. *Source* Hens et al. (2010)

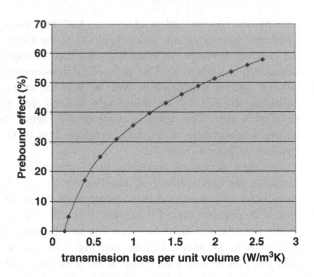

same EPC rating, in MWh/dw.y, as a small thermally inefficient dwelling. However, the comparison is interesting.

These researchers plot the EPC against actual fuel consumption. Similar to the German datasets examined above, the median actual consumption is 40% below the EPC; the percentage gap increases as the EPC increases; and this gap becomes zero or negative for highly energy-efficient homes, i.e. with very low EPC.

The authors also relate the EPC to the proportion of household income spent on heating. They graph this against 'energy intensity', equivalent to 'heating factor' in Tighelaar and Menkveld (2011). Their results show that on average most households in their dataset spend 2–5% of their income on space heating and achieve an energy intensity of around 0.6 (a prebound effect of 40%). In cases where income is low or the EPC is high, some households fail to achieve an energy intensity as high as 0.5, even by spending up to 7.5% of their income on space heating. Households in buildings with lower consumption, or who have higher incomes, achieve an energy intensity of 0.8–1.1 (a prebound effect of 20% to − 10%) spending less than 2% of their income on space heating.

Here, we are seeing a relationship between income and a form of the prebound effect. In general, people do not seem to want to spend more than around 5% of their income on heating, so if their income is low or their house requires a lot of energy to heat (either because it is large or of poor thermal quality), they tend to under-heat, with high prebound effects. If their income is high or their home is thermally less demanding, they tend to overheat, with low, zero or negative prebound effects. This supports the hypothesis put forward in Sect. 5.4, that the most likely factor influencing the way the prebound effect varies with Q_P is the behavior of the occupants.

5.6 Conclusions and Implications

In independently gathered datasets, we found a consistent set of relationships between the calculated heating energy consumption of German dwellings (Q_P) and the actual measured consumption (Q_M). First, for any particular value of Q_P, there is a large range of values of Q_M. This is not surprising, since different households use similar dwellings very differently. Second, Q_M is, on average, 30% below Q_P. This result that can partially be explained by the methodology used to calculate Q_P, which assumes high year-round indoor temperatures throughout the whole dwelling, with excessive ventilation. However, it could also be the case that, on average, German householders today are more thrifty with heating energy than was assumed at the time the methodology was developed. Third, the percentage gap between Q_P and average Q_M increases as Q_P increases, and fourth, it falls to zero for homes with Q_P of around 100 kWh/m^{2}a and goes negative thereafter (except, apparently, and for passive houses). These last two findings cannot be explained by shortcomings in the methodology for calculating Q_P. The most likely explanation is that, on average, householders are constraining their heating behavior to limit their fuel consumption if their homes are thermally leaky, but heating more liberally if their homes are energy efficient.

Studies of datasets in the Netherlands, the UK, Belgium, and France showed similar tendencies. Graphs of variables corresponding to Q_P and Q_M, in the Netherlands and Belgium, showed the same pattern of variation between these two variables as in the German studies, and one German study claimed to have found similar tendencies in Danish datasets. Data from France showed a comparable effect, though here the dependent variable conflated dwelling size with thermal quality. This suggests the need for further studies in these and other countries, especially with similar climates and building types to see how general these patterns are in household heating consumption.

Using curve fitting, we were able to offer a basic graphical model of the prebound effect. Plots of Q_M against Q_P also showed that for Q_P above about 120 kWh/m^{2}a, on average, each 1% increase in calculated energy consumption (Q_P) is associated with a 0.2–0.5% increase in actual consumption (Q_M).

For Germany we can conclude that using Q_P to predict heating energy consumption is a mistaken endeavor, as actual consumption bears little relation to it for any specific home. More interestingly, the current German practice of using Q_P to calculate energy savings through thermal renovation is also a serious mistake. To begin with, Q_P tells us virtually nothing about the actual energy consumption of a particular home pre-or post-retrofit, since there is so much variation in Q_M for any particular value of Q_P. Hence, a homeowner may get highly misleading estimates of fuel saving, through renovation, if she bases her calculations on Q_P. Government ministries and agencies need to make this point clear in their promotional material as they seek to persuade homeowners to thermally retrofit their homes.

Second, the 30% gap between average Q_P and average Q_M indicate that there is far less savings potential in the German housing stock than is commonly claimed. German households cannot save fuel they are not already consuming. As we saw in Chap. 2, evaluations of Germany's more ambitious retrofit projects—those which push beyond what the law requires, with the aid of Federal subsidies—are saving an average of around 33%, based on pre- and post-retrofit values of Q_P (Clausnitzer et al. 2009). If we base savings estimates instead on real consumption, Q_M, the savings fall to around 25%. Our modeling above suggested that even where the calculated savings are as high as 55%, real savings may be only 37%, due to the disparity between Q_P and average Q_M.

Finally, the prebound effect suggests there might be a rich avenue for heating fuel savings that policymakers have hardly begun to explore. The gap between Q_P and average Q_M suggests that German householders are, on average, far more fuel thrifty than they are often given credit for. Further, the large range of Q_M for each value of Q_P suggests that some households are learning to live quite comfortably with an even lower heating intensity than the average German household. Some of this may well be symptomatic of fuel poverty, but there appear to be many households where deep fuel savings are made, through behavioral change, that do not lead to discomfort. A handful of research projects over the last 15 years have revealed the types of strategies some households use, to reduce their heating bills without losing thermal comfort (e.g. Martiskainen 2008; UBA 2006). At least one study has taken this further, asking what motivates some householders to adopt such strategies while others make less effort to do so (Gram-Hanssen 2010). The prevalence of the prebound effect in German home heating suggests that this type of research could make a significant contribution to fuel savings in the future.

References

Barker T, Ekins P, Foxton T (2007) The macro-economic rebound effect and the UK economy. Energy Policy 35(10):4935–4946

Beecken C, Schulze S (2011) Energieeffizienz von Wohngebäuden: Energieverbräuche und Investitionskosten energetischer Gebäudestandards. Bauphysik 33(6):338–344

Berndgen-Kaiser A, Fox-Kämper R, Holtmann S, Frey T (2007) Leben im Passivhaus: Baukonstruktion, Baukosten, Energieverbrauch, Bewohnererfahrungen, Institut für Landes- und Stadtentwicklungsforschung und Bauwesen des Landes Nordrhein-Westfalen, ILS NRW, Aachen, 68 S.-ILS-NRW-Schriften Bd. 202. http://www.ils-forschung.de/index.php?option =com_content&view=article&id=348&Itemid=205&lang=de. Accessed 18 Nov 2011

BMVBS (Bundesministerium für Verkehr, Bau und Stadtentwicklung (2011) Wohnen und Bauen in Zahlen 2010/2011, 6. Auflage, Stand: Mai, 2011. http://www.bmvbs.de/cae/servlet/ contentblob/54084/publicationFile/25231/wohnen-und-bauen-in-zahlen-2009-2010.pdf. Accessed 28 March 2012

Brookes L (1979) A low-energy strategy for the UK by G. Leach et al.: a review and reply. Atom 269:3–8

Cayla J-M (2010) From practices to behaviors: estimating the impact of household behavior on space heating energy consumption. In: Proceedings of ACEEE summer study on energy efficiency in buildings, Pacific Grove, CA, 15–20 Aug 2010

Cayre E, Allibe B, Laurent M-H, Osso D (2011) There are people in the house! How the results of purely technical analysis of residential energy consumption are misleading for energy policies. In: Proceedings of ECEEE 2011 summer study, energy efficiency first: the foundation of a low-carbon society, Belambra Presqu'ile de Giens, 6–11 June 2011. ECEEE, Stockholm

Clausnitzer KD, Gabriel J, Diefenbach N, Loga T, Wosniok W (2009) Effekte des CO_2-Gebäudesanierungsprogramms 2008. Bremer Energie Institut, Bremen

DIN (2003a) DIN 4108–2:2003–07 Wärmeschutz und Energie-Einsparung in Gebäuden: Mindestanforderungen an den Wärmeschutz. Verlag, Berlin

DIN (2003b) DIN EN 832:2003-06 Wärmetechnisches Verhalten von Gebäuden: Berechnung des Heizenergiebedarfs—Wohngebäude. Verlag, Berlin

Enseling A, Hinz E (2006) Energetische Gebäudesanierung und Wirtschaftlichkeit—Eine Untersuchung am Beispiel des 'Brunckviertels' in Ludwigshafen. Institut Wohnen und Umwelt, Darmstadt

Erhorn H (2007) Bedarf-Verbrauch: Ein Reizthema ohne Ende oder die Chance für sachliche Energieberatung? Fraunhofer-Institut für Bauphysik, Stuttgart. http://www.buildup.eu/publications/1810. Accessed 20 Nov 2011

Galvin R (2012) German federal policy on thermal renovation of existing homes: a policy evaluation. Sustain Cities Soc 4:58–66

Gram-Hanssen K (2010) Residential heat comfort practices: understanding users. Building Res Inf 38(2):175–186

Gram-Hanssen K (2011) Households' energy use—which is the more important: efficient technologies or user practices? In: Proceedings of world renewable energy congress 2011, Linköping, 8–13 May 2011

Greller M, Schröder F, Hundt V, Mundry B, Papert O (2010) Universelle Energiekennzahlen für Deutschland—Teil 2: Verbrauchskennzahlentwicklung nach Baualtersklassen. Bauphysik 32(1):1–6

Guerra Santin O, Itard L, Visscher H (2009) The effect of occupancy and building characteristics on energy use for space and water heating in Dutch residential stock. Energy Buildings 41:1223–1232

Haas R, Biermayr P (2000) The rebound effect for space heating: empirical evidence from Austria. Energy Policy 28:403–410

Hens H, Parijs W, Deurinck M (2010) Energy consumption for heating and rebound effects. Energy Buildings 42:105–110

Holm SO, Englund G (2009) Increased ecoefficiency and gross rebound effect: evidence from USA and six European countries 1960–2002. Ecol Econ 68:879–887

ILS (Institut für Landes- und Stadtentwicklungsforschung) (2010) Leben im Passivhaus: Baukonstruktion, Baukosten, Energieverbrauch, Bewohnererfahrungen. ILS, Dortmund

Jagnow K, Wolf D (2008) Technische Optimierung und Energieeinsparung. OPTIMUS, Hamburg City-State

Jakob M (2006) Marginal costs and co-benefits of energy efficiency investments: the case of the Swiss residential Sector. Energy Policy 34:172–187

Jakob M (2007) The drivers of and barriers to energy efficiency in renovation decisions of single-family home-owners. CEPE Working paper No. 56. Centre for energy policy and economics, Zurich

Jevons S (1865) The coal question—can Britain survive?. Macmillan and Co, London

Kaßner R, Wilkens M, Wenzel W, Ortjohan J (2010) Online—Monitoring zur Sicherstellung energetischer Zielwerte in der Baupraxis, Paper for 3. Effizienz Tagung Bauen + Modern-isieren, 19–20 November 2010, Hannover. http://www.energy-check.de/wp-content/uploads/2010/11/EFT_2010_ortjohann_2010-10-18.pdf. Accessed 18 Nov 2011

Khazzoom JD (1980) Economic implications of mandated efficiency standards for household appliances. Energy J 1:21–40

Kelly S (2011) Do homes that are more energy efficient consume lessenergy?: A structural equation model for England's residential sector. Energy 36(9):5610–5620

Knissel J, Loga T (2006) Vereinfachte Ermittlung von Primärenergiekennwerten. Bauphysik 28(4):270–277

Knissel J, Alles R, Born R, Loga T, Müller K, Stercz V (2006) Vereinfachte Ermittlung von Primärenergiekennwerten zur Bewertung der wärmetechnischen Beschaffenheit in ökologischen Mietspiegeln. Deutsche Bundesstiftung Umwelt, Osnabrück

Loga T, Diefenbach N, Born R (2011) Deutsche Gebäudetypologie. Beispielhafte Maßnahmen zur Verbesserung der Energieeffizienz von typischen Wohngebäuden. Institut Wohnen und Umwelt, Darmstadt

Maaß JB, Walther C, Peters I (2008) Erfahrungen mit Passivhaussiedlungen in Deutschland (Schwerpunkt Norddeutschland). Bezirksamt Harburg, Hamburg

Martiskainen M (2008) Household energy consumption and behavioural change—the UK perspective. In: T.G. Ken, A. Tukker and C. Vezzoli (eds) Sustainable consumption and production: framework for action, proceedings of the 2nd conference of the sustainable consumption research exchange (SCORE!) network, Flemish Institute for Technological Research, Mol, pp. 73–90

Peper S, Feist W (2008) Gebäudesanierung „Passivhaus im Bestand" in Ludwigshafen/Mundenheim Messung und Beurteilung der energetischen Sanierungserfolge. Passivhaus Institut, Darmstadt

Roth K, Engelman P (2010) Impact of user behavior on energy consumption in high-performance buildings—results from two case studies. Presentation at Fraunhofer Center for Sustainable Energy Studies, Denver

Schloman B, Ziesling HJ, Herzog T, Broeske U, Kaltschnitt M, Geiger B (2004) Energieverbrauch der privaten Haushalte und des Sektors Gewerbe, Handel, Dienstleistungen (GHD), Projekt Nr 17/10, Abschlussbericht an das Ministerium für Wirtschaft und Arbeit. Fraunhofer Institut für Systemtechnik und Innovationsforschug, Karlsruhe. http://isi.fraunhofer.de/isi-de/e/projekte/122s.php. Accessed 26 Nov 2011

Schröder F, Engler HJ, Boegelein T, Ohlwärter C (2010) Spezifischer Heizenergieverbrauch und Temperaturverteilungen in Mehrfamilienhäusern—Rückwirkung des Sanierungsstandes auf den Heizenergieverbrauch. HLH 61(11):22–25. http://www.brunata-metrona.de/fileadmin/Downloads/Muenchen/HLH_11-2010.pdf. Accessed 8 Dec 2011

Schröder F, Altendorf L, Greller M, Boegelein T (2011) Universelle Energiekennzahlen für Deutschland: Teil 4: Spezifischer Heizenergieverbrauch kleiner Wohnhäuser und Verbrauchshochrechnung für den Gesamtwohnungsbestand. Bauphyisk 33(4):243–253

Simons H (2012) Energetische Sanierung von Ein- und Zweifamilienhäusern Energetischer Zustand Sanierungsfortschritte und politische Instrumente. Bericht im Auftrag des Verbandes der Privaten Bausparkassen e.V. Empirica, Berlin

Sorrell S, Dimitropoulos S (2008) The rebound effect: microeconomic definitions, limitations and extensions. Ecol Econ 65(3):636–649

Stern PC (2000) Towards a coherent theory of environmentally significant behaviour. J Soc Issues 56:407–424

Sunikka-Blank M, Galvin R (2012) Introducing the prebound effect: the gap between performance and actual energy consumption. Building Res Inf 40:260–273

Thomsen K, Schultz J, Poel B (2005) Measured performance of 12 demonstration projects—IEA Task 13 'advanced solar low energy buildings'. Energ Buildings 37:111–119

Tighelaar C, Menkveld M (2011) Obligations in the existing housing stock: who pays the bill? In: Proceedings of the ECEEE 2011 summer study, energy efficiency first: the foundation of a low-carbon society, Belambra Presqu'ile de Giens, 6–11 June 2011. ECEEE, Stockholm

UBA (Umwelt Bundesamt) (2006) Wie private Haushalte die Umwelt nutzen –höherer Energieverbrauch trotz Effizienzsteigerungen. UBA, Dessau

Walberg D, Holz A, Gniechwitz T, Schulze T (2011) Wohnungsbau in Deutschland—2011 Modernisierung oder Bestandsersatz: Studie zum Zustand und der Zukunftsfähigkeit des deutschen „Kleinen Wohnungsbaus", Arbeitsgemeinschaft für zeitgemäßes Bauen, eV

Chapter 6
The Economics of Thermal Retrofits in Germany

Abstract German law restricts the thermal demands of building regulations to levels that are 'economically viable' (*wirtschaftlich*), i.e. that pay back, through fuel savings, within the technical lifetime of the thermal measures. In an effort to reduce CO_2 emissions deeply and rapidly, the government sets thermal regulations as tightly as this law allows. But the cost-benefit models, which its estimates are based on, have an uneasy relationship with the realities of German buildings and their owners: they take account only of retrofit costs that are directly do to with thermal features; they do not allow for difficult physical features of actual German residential buildings; they treat homeowners as commercial investors with long time horizons for the return on capital investment; and they are based on calculated, rather than actual, pre-retrofit consumption. This makes the claim of economic viability questionable in the eyes of many homeowners. The models also demand levels of investment that lead to costs of CO_2 abatement 10–30 times higher than is currently achieved in other sectors. We argue that the government needs to revise its focus on economic viability and consider, instead, an emphasis on what is economically optimum. It also needs to take advantage of the significant contribution that household behavior is already making to heating energy consumption savings.

Keywords Economic viability · Thermal retrofit · Payback time · CO_2 abatement · Green Deal

6.1 Introduction

A key to understanding both the appeal of thermal retrofits, and the difficulties Germany's retrofit programme are experiencing, lies in the concept of 'economic viability'. As we saw in Chap. 2, the thermal measures demanded in the Energy Saving Regulations (EnEV) are claimed to be 'economically viable'. This means they are set at such a level that the fuel savings they are expected to bring, over their technical lifetime, will pay back all the money invested in them. This applies

R. Galvin and M. Sunikka-Blank, *A Critical Appraisal of Germany's Thermal Retrofit Policy*, Green Energy and Technology, DOI: 10.1007/978-1-4471-5367-2_6, © Springer-Verlag London 2013

to both thermal upgrades on existing homes, and the thermal aspects of new builds. In both cases, the value of the fuel expected to be saved has to cover the difference between a design without thermal quality and a design to EnEV standards.

In Germany, a legal concept lies behind economic viability. The standards demanded by the EnEV are limited by the Energy Saving Law (EnEG—*Energieeinsparungsgesetz*), which states that the demands of the EnEV:

> ... must be economically viable (*'wirtschaftlich vertretbar'*) using current technology. Demands qualify as economically viable when, in general, the measures demanded pay back, through the savings they bring, within the usual useful lifetime of the thermal measures (EnEG, para. 5, authors' translation).

When each new version of the EnEV is drawn up, the Housing Ministry (BMVBS) has to prove, to the satisfaction of both houses of the Federal Parliament, that the demands it is making fit this criterion. To achieve this, it commissions an expert report before each upgrade of the EnEV, where a mathematical model for calculating payback time is set alongside current costs of typical thermal upgrade measures, showing how these are economically viable. The expert reports for EnEV 2002 (Feist 1998) and EnEV 2009 (Kah et al. 2008) were both prepared by the Passive House Institute, and supported the 30% tightening of thermal standards on each occasion. Up until late 2011, the Housing Ministry was planning a further tightening of 30% for September 2012. By this time, however, severe doubts were being expressed by a number of professionals, academics, and practitioners, including a new report commissioned by the Housing Ministry (Hauser et al. 2012). Our discussions with Federal MPs involved in the EnEV update over this period indicated that the concerns of these voices were gradually being heeded, and in July 2012 the Housing Minister announced there would be no tightening of standards in 2012 (BMVBS 2012).

Economic viability is a legal concept designed to protect the interests of homeowners who are required, by the regulations, to include thermal measures in renovations or repairs to their homes. But it also functions as a key policy plank to promote thermal renovation. Government ministries and agencies publish leaflets, hold seminars, and produce websites seeking to persuade homeowners to thermally retrofit their properties on the grounds that this will not cost them any money in the long run. Thermal renovation 'pays for itself' (*es lohnt sich*) declares a government sponsored national newspaper advertisement (Bundesregierung 2012). Thermal retrofitting of detached houses 'pays' (*rechnet sich*) says a headline from the German Energy Agency (2012a). 'Thermal renovation is economically viable,' declares the headline of an article from the same Agency setting out to rebut critics (2012b). The German Development Bank (*Kreditanstalt für Wiederaufbau*—KfW) produces detailed arguments to prove that thermal retrofitting not only 'pays for itself', but is 'one of the most economical forms of climate protection'. Our discussions with policy actors at Federal, state and municipal level over recent years, and our reading of their published material, indicate that the notion that thermal retrofitting to EnEV standards always pays back is the main promotional lever used to persuade homeowners to undertake it.

This chapter explores the notion of economic viability of thermal retrofits in the German context, drawing out implications that may be of more general interest for other countries. Section 6.2 explains how and why only certain costs are included in the calculation of economic viability. Section 6.3 explores how certain peculiarities of existing buildings can compromise the economic viability of apparently standard thermal retrofits. Section 6.4 looks at the basics of mathematical models used to assess economic viability, focusing on a quirk in these that can cause unhappy surprises for retrofitting homeowners. Section 6.5 explores how the prebound and rebound effects interfere with standard notions of economic viability. Section 6.6 shows how 'economically viable' differs from 'economically efficient' and 'economically optimum' in assessing returns on investment, and how this bears on cost-effective CO_2 abatement. Section 6.7 explores how the condition of 'economic viability' might work in the UK context, and Sect. 6.8 offers conclusions and recommendations for policymakers.

6.2 Which Costs are Counted

German policymakers and their expert advisors make a careful distinction between 'anyway costs' (*sowieso-Kosten*) and 'additional thermal costs' (*energetische Mehrkosten* or *energiebedingte Mehrkosten*). If, for example, a homeowner needs to replace a window for the reason that it is no longer air tight, she is required by the EnEV to install a high thermal quality model, of U-value 1.30 W/m^{2}K or better (EnEV 2009). Typically, this will cost around €800, but the value of the heating fuel savings it brings over its 25-year lifetime will be far <€800. For example, if the window area represents 0.5% of the building envelope (1 m^2 out of 200 m^2) and the annual heating fuel bill is the German average of around €1000, the window will save roughly €100 during its lifetime.

However, German policymakers argue that this is a false comparison, because the old window had to be replaced anyway (Simons 2012). Therefore, the homeowner should not count its full cost, but only the *additional* cost due to its high thermal quality, usually rated at around 10% of its total value. Hence the 'additional thermal costs' were only €80, and therefore the replacement *was* economically viable, because the investment of €80 brought a return of €100.

This approach is used for all thermal upgrade measures. For external wall insulation, we may only count the polystyrene blocks and the labor of fixing them to the wall, not the costs of scaffolding or of cutting away the old render and applying new render and paint: the wall was old and needed a new render and paint anyway. For roof insulation, we do not count the costs of laying tiles over the insulation slabs we are inserting, as the existing tiles were old and due for replacement anyway. We count only a portion of the cost of a new, energy-efficient boiler, because the previous one was old and nearing the end of its useful life. If a ventilation system has to be installed to compensate for the new air tightness of the house, we do not count its cost at all, as it adds to the value of the house by improving its indoor air quality. Possibly the only upgrade measure that gets counted in full is insulation added to the basement ceiling, which is entirely for thermal reasons and serves no other purpose.

Hence, the costs that are included in the calculation of economic viability are much lower than the total costs of the retrofit. In projects we have observed in detail, the full costs for a comprehensive retrofit to EnEV standards may range from €500 to €1,200 per m² of floor area, while the 'additional thermal costs' may range from €100 to €300 (cf. Simons 2012; but compare to Enseling and Hinz 2006).

The author of the expert report on economic viability for the first version of the EnEV rejects criticism that this produces artificially deflated cost figures for thermal retrofits. Speaking at the UN Conference on Energy and Housing in Vienna in November 2009, Feist (2009) emphasized that his calculations for economic viability were only ever intended to apply to buildings that were due for major maintenance and would have to be renovated anyway. These cases, he maintained, would always be economically viable, because the only costs that needed to be paid back through fuel savings were the additional thermal costs.

While this approach might seem like bookkeeping sleight-of-hand to some homeowners, it is becoming widely accepted internationally. The additional thermal costs are the only element considered in economic viability and payback time calculation models in a raft of recent peer-reviewed studies on economic modeling of thermal retrofits (e.g., Kumbaroğlu and Madlener 2012; Martinaitis et al. 2004, 2007; Tuominen et al. 2012).

Of course, this creates a problem for homeowners who do not believe their properties are due for major maintenance but who want to thermally retrofit to keep warmer or protect the environment. They have to upgrade to EnEV standards even though all the costs are, for them, additional thermal costs.

If homeowners have to wait until they believe their properties are due for major maintenance, many will be waiting a very long time. This could be one reason the rate of thermal renovation is so far behind what policymakers planned for. But even for those whose properties are due for comprehensive maintenance or repair, there can be difficulties with economic viability. These have to do with the peculiarities of existing buildings, an issue we now turn to discuss.

6.3 Technical Limitations and the Economics of Thermal Renovation

In Chap. 4, Sect. 4.4 we explored some of the technical problems with thermal renovation to the high standards demanded by the EnEV, due to buildings' peculiarities. One practical example was the 'wall and roof dilemma', where the roof overhang has to be extended to accommodate the 12–16 cm of insulation required on the wall. Also, the roof itself might have to be raised to accommodate its required 20–22 cm of insulation and to avoid thermal bridges at the wall-roof joins (see Figs. 4.4 and 4.5). Further, air-tight window frames cause indoor air quality problems which are tiresome to solve: the options are a regular, managed household ventilation routine; an expensive whole-house ventilation system with heat recovery; or constant trickle ventilation which wastes energy and increases fuel bills (Galvin 2013).

These problems are particularly evident in 'small' residential buildings, i.e. those with 1–6 dwellings, since these have a high ratio of corner, wall and roof joins compared to plain and problem-free areas of wall or roof. This sector includes most owner-occupied houses and privately rented apartments. Our discussions with the German Energy Agency indicate that this sector is by far the slowest to take up thermal renovation. Ulrike Hacke, a sociologist at the Institut Wohnen und Umwelt (www.iwu.de) who researches landlord-tenant relationships, writes:

> Because lawmakers have continually raised the legal minimum standards for energy efficiency of dwellings in recent years, housing investors have spoken increasingly of the economic viability problems this brings. This could lead to a negative effect on investment. (Hacker 2009, p. 1, authors' translation)

There is, nevertheless, a possible solution for such cases. Clause 25 of the EnEV provides that homeowners can apply to their local authority for an 'exception' (*Ausnahme*) if retrofitting to EnEV standards would be of *besonderer Umstand* (special circumstance), *unangemessener Aufwand* (disproportionate expense), or *unbillige Härte* (inequitable hardship). Elements within the construction industry have developed the skills to make good use of this clause. For example, an in-house publication of the *Bundesverband Farbe Gestaltung Bautenschutz* (German Federal Association for Paint Design and Preservation of Structures, www.farbe.de), labeled 'Exclusively for Guild Members,' gives specific, detailed advice on how to make the best use of it.

For individuals the application process is onerous, and the applicant risks a refusal of all or part of his request. Due to the lack of inspection, in many cases homeowners have simply ignored the EnEV standards and retrofitted to what they found sensible. But even for homeowners whose properties are due for maintenance, and do not suffer any of these technical limitations, economic viability may be complicated due to the timing of the savings from heating fuel reductions, an issue we discuss next.

6.4 The Timing of the Savings: The Issue of Exponential Curves

For the ideal case of a problem-free building that is due for major or comprehensive maintenance, a thermal retrofit is considered economically viable if the total monetary savings expected to be gained from fuel savings are equal to or greater than the additional thermal costs. This can be expressed as:

$C \leq B$, where
C = additional thermal costs;
B = monetary benefit, i.e. money saved through reduced fuel costs over the technical lifetime of the thermal renovation measures.

Considering, first, the monetary benefit, this is worked out using a formula of the type:

$$B = Q \times P_1 \times \frac{A^N - 1}{A - 1}, \text{ where}$$

Q = Quantity of fuel saved each year (kWh)
P_1 = price of each unit of fuel saved in year 1 (€/kWh)
N = number of years of the technical lifetime of the renovation measures
A = an 'annuity factor'[1]

The 'annuity factor' is a composite figure, usually close to 1.0, made up of the expected annual fuel price rise and the 'discount rate', i.e. the losses incurred in having to wait until future years to get the payback. Suppose, for example, the expected annual fuel price rise is 4%, the variable A would be 1.04. If the losses amount to, say, 2% for every year we have to wait for the savings, the variable A would now be 1.04/1.02. Discount rates are a thorny issue and we will leave them to later in the discussion.

We can use the model above for 'real' price increases, i.e. over-and above inflation, or 'nominal' price increases, i.e. including inflation. Either way is correct, as long as we are consistent with C, the additional thermal costs. Just for now we will work with 'nominal' prices and costs.

Suppose we take out a table mortgage to pay the additional thermal costs, and our 'nominal' interest rate is 2%. If the additional thermal costs are €30,000 and N = 25 years, our monthly repayments will be €127 (calculated using a standard table mortgage formula), and C will be €38,100 (additional thermal costs plus interest on the loan). For the retrofit to be economically viable, the *average* monthly fuel saving would therefore have to be at least €127.

Suppose the retrofit is just economically viable, so that the average monthly fuel saving is, indeed, €127. The problem now arises that the homeowner will not, actually, start saving €127 per month on fuel bills until 14 years after the retrofit. This is because the fuel savings rise as an exponential curve, based on a 4% annual fuel price increase, while the payments are constant at €127 per month. To achieve an average value of €127 over 25 years, savings rising at this rate would actually begin at €76 per month. In the 14th year they would reach €127 per month, rising to €195 per month in the 25th year. Figure 6.1 displays this on a graph.

---

[1] The expression $(A^N - 1)/(A - 1)$ is the sum of the geometric sequence $A^0 + A^1 + A^2 + \ldots$ A^{N-1}, i.e. it adds up the savings made for all the years of the technical lifetime of the retrofit measures. A fuller explanation will be given of this in Chap. 8.

Fig. 6.1 Monthly fuel savings and loan payments for thermal retrofit that is just economically viable. 'Additional thermal costs' of retrofit are €30,000, taken as a table mortgage at real interest rate 2%; expected nominal annual fuel price rise is 4%. The monthly savings reach the level of monthly payments in the 14th year

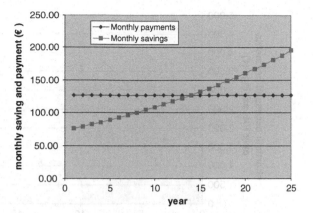

The homeowner, then, suffers an ever-deepening shortfall for the first 14 years. This gradually reduces after the 14th year until the end of the 25th year, when it falls to zero. Figure 6.2 displays this on a graph, showing how the cumulative shortfall grows to €4,602 in the 14th year, before gradually falling to zero.

Although the homeowner is genuinely getting her money back over the 25-year lifetime of the thermal retrofit measures, she may be surprised to find that the value of her fuel savings is only half the cost of her loan payments in the first year. She may be disappointed that it takes 14 years for the two to draw level, and that she has to wait yet another 11 years to recoup the losses she has incurred up to that time.

The only way to avoid this problem is to assume a discount rate at least equal to the annual fuel price rise. Then the annual fuel price rise is canceled out by the discount rate (e.g. $A = 1.04/1.04 = 1.0$). In practice, this means that the value to an investor of a fuel price rise in, say 10 years time, is canceled out by the fact that he is forced to wait 10 years to get the money. A discount rate of 4% would flatten out the effective fuel savings to €76 every month (the same amount as the first month's saving) for the 25-year period, producing a total saving of €22,800 rather than €38,100. With a mortgage interest rate of 2% this means the homeowner could only borrow €17,930 rather than €30,000, severely limiting the depth of retrofitting he could do, so that he is far less likely to reach the target of €76 in fuel savings per month. If the discount rate is set higher than the expected fuel price rise (UK construction firms often set it at 8%), the value of fuel savings falls each year, rather than rising, so the amount the investor can afford to spend on the retrofit, to make it economically viable, must be revised further downward. The calculations are affected by the fact that German policy actors use low discount rates, generally around 5% 'nominal' or 3% 'real' (e.g. Enseling and Hinz 2006; Kah et al. 2008). Hence, the cost and benefit curves tend to be shaped as in Figs. 6.1 and 6.2, and this is one of the reasons homeowners are often disappointed that their immediate returns are less than expected.

These issues aside, a further question concerns the amount of fuel actually saved through a thermal retrofit, an issue we will now discuss.

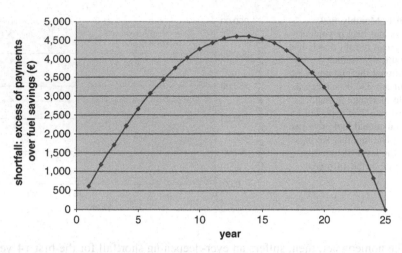

Fig. 6.2 Accumulated shortfall as total excess of payments over fuel savings for thermal retrofit that is just economically viable. 'Additional thermal costs' of retrofit are €30,000, taken as a table mortgage at nominal interest rate 2%; expected nominal annual fuel price rise is 4%. The shortfall rises to a maximum in the 14th year, then diminishes, falling to zero at the end of the 25th year

6.5 Prebound Effect and Economic Viability

In Chap. 5, we explored the ubiquitous issue of the discrepancy between the calculated and the actual, measured thermal performance of German homes.

We found that, on average, measured heating energy consumption is 30% lower than calculated consumption in Germany, and that this gap widens as the thermal quality of dwellings falls. We called this the 'prebound effect', as it is the opposite of the rebound effect, where a dwelling consumes more than its calculated rating (and see Sunikka-Blank and Galvin 2012). We found that the rebound effect is evident in highly energy-efficient buildings, while the prebound effect dominates in the larger part of the housing stock.

We suggested that this skews calculations of the energy expected to be saved through retrofits. If we think a dwelling is consuming, say, the national average calculated rating of 225 kWh/m²a, and we design a retrofit strategy to reduce this to 100 kWh/m²a, we will credit our project with fuel savings of 125 kWh/m²a. If this is an average sized German dwelling of living area 87 m² ('useful' area 105 m²), the annual fuel savings will be 13,125 kWh, worth €1,313 in the first year, or €109 per month, assuming a natural gas price of €0.10/kWh.

But if this is an average dwelling it may not be consuming 225 kWh/m²a, but more likely only around 156 kWh/m²a, due to the prebound effect. If our retrofit measures reduce its consumption to 100 kWh/m²a, the actual annual savings will be 5,824 kWh, worth the lesser sum of €583, or €49 per month.

On paper our project may well be economically viable. At an annual interest rate of 2%, steady monthly fuel savings worth €109 (assuming a discount rate of 4%) could pay off a loan of €25,7172 over 25 years. However, if the pre-retrofit *actual* consumption was the lower figure of 156 kWh/m²a, the retrofit will bring savings worth only €49 per month and we will suffer a loss each month of €60, totalling €18,000 over the 25-year period. Figure 6.3 illustrates this type of situation schematically.

This puts a further question mark over the usefulness of the concept of 'economic viability' as it is framed in German policy and practice. Many projects may appear economically viable if we use only theoretical, calculated figures for their pre- and post-retrofit consumptions. But in real life, homeowners' budgets are affected by actual fuel saved, not theoretical fuel quantities. In Chap. 2, we saw that the calculated savings in Germany's more ambitious thermal retrofit projects are achieving average heating fuel savings of around 33%, based on national surveys by Clausnitzer et al. (2009, 2010) and Diefenbach et al. (2011). When the prebound effect is taken into consideration this falls to 25%. Despite its protestations that 'thermal retrofits are economically viable' (see above), the German Energy Agency now seems to accept that actual savings averaging only about 25% are being achieved, in practice (DENA 2012c). It is difficult to see how these would be 'economically viable' if designed to achieve the exact standards of the EnEV.

A recent empirical study lends support to these figures. Michelsen and Müller-Michelsen (2010) investigated the economic viability of thermal retrofits in the

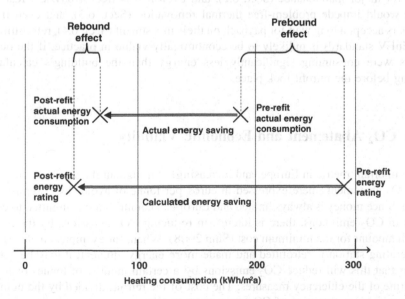

Fig. 6.3 Schematic showing the effect of the prebound and rebound effect on the amount of energy saved through a thermal retrofit

German housing stock, using data from the Institut Wohnen und Umwelt. The data cover 156,866 buildings, each with two or more dwellings, and are grouped according to the age of the buildings. The heating energy consumption values given in the data are based on actual consumption, called *'verbrauchsbasierter Energiekennwert'* (usage-based energy rating) in the study. Consumption ranges from a high of 162 kWh/m²a for smaller (1–6 dwellings) buildings erected in 1919–1948, to a low of 123 kWh/m²a for the largest buildings erected in 1984–1994. Table 6.1 presents estimates of the thermal standard to which, these authors calculate, each of the classes of building could be retrofitted in an economically viable thermal renovation project based on current construction costs.

The study finds that the largest economically reliable reduction is 41%, for large apartment blocks built between 1958 and 1968. Most other categories offer gains of around 20%. The smallest buildings erected since 1979 offer <10%. Leaving aside the post-1995 buildings (already relatively efficient and less likely to be renovated), the average gain is around 22%. The average for post-war housing (1949–1979) is 25.4%.

Further, policy actors often say that an important benefit of thermal renovation is increased comfort indoors: a warmer home. But this does not square with the German model of economic viability. The heating consumption figures of both pre- and post-retrofit consumptions are based on the theoretical model of DIN 4108 (see Sects. 4.3 and 5.3), which assumes standardized indoor heating behavior. The economic viability calculations *only* work if identical indoor temperature regimes are used for both pre- and post-retrofit scenarios.

Taking the prebound effect into account suggests, then, that even if a building is due for major maintenance (Sect. 6.2), and even if it is free of physical features that would impede problem-free thermal renovation (Sect. 6.3), and even if its owners accept a long wait for payback on their investment (Sect. 6.4), retrofitting it to EnEV standards is unlikely to be economically viable in practice, if the occupiers were consuming significantly less energy than the building's calculated rating before the retrofit took place.

6.6 CO₂ Abatement and Economic Viability

An important metric in Europe, and increasingly throughout the world, is the 'cost of CO_2 abatement', here expressed in euros per tonne of avoided CO_2 emissions (€/t). Since money is always limited and European countries are committed to deep cuts in CO_2 emissions, there is interest in reducing CO_2 emissions by the maximum amount for the minimum cost (Sinn 2008). When, for example, an electricity generating station is retrofitted and made more energy efficient, it may be calculated that this will reduce CO_2 emissions by a certain number of tonnes over the lifetime of the efficiency measures. The cost of the retrofit divided by the number of tonnes gives the price of CO_2 abatement, in €/t.

Table 6.1 Comparison of actual heat energy consumption and estimated consumption to which each class of building could be economically viably renovated, with percentage estimated gain given (given in, and calculated from, Michelsen and Müller-Michelsen 2010, p. 454). Abbreviation HE = heating energy

Number of dwellings in a building	Year of building	Actual HE consumption (kWh/m^{2}a)	Potential HE consumption (kWh/m^{2}a)	Potential economically viable HE saving (kWh/m^{2}a)	Percentage potential fuel saving
2–6	1900–1918	159	137	22	13.8
7–12		141	126	15	10.6
13–21		140.5	120	20.5	14.6
>21		135	122	13	9.6
2–6	1919–1848	162	136	26	16.0
7–12		152	118	34	22.4
13–21		141.5	109	32.5	23.0
>21		141	104	37	26.2
2–6	1949–1957	160	134	26	16.3
7–12		148	116	32	21.6
13–21		134	106	28	20.9
>21		126	91	35	27.8
2–6	1958–1968	161	128	33	20.5
7–12		150	110	40	26.7
13–21		150	100	50	33.3
>21		144	85	59	41.0
2–6	1969–1978	151	131	20	13.2
7–12		146	121.5	24.5	16.8
13–21		151	105	46	30.5
>21		140	89	51	36.4
2–6	1979–1983	143	137	6	4.2
7–12		137	117	20	14.6
13–21		136	102.5	33.5	24.6
>21		134	90	44	32.8
2–6	1984–1994	136	125	11	8.1
7–12		133	112	21	15.8
13–21		131	91	40	30.5
>21		123	82	41	33.3
2–6	1995 on	108	108	0	0.0
7–12		103	103	0	0.0
13–21		101	101	0	0.0
>21		98	98	0	0.0
Averages		**138.0**	**111.1**	**26.9**	**18.9**
Averages pre-1995		**143.1**	**112.3**	**30.8**	**21.6**

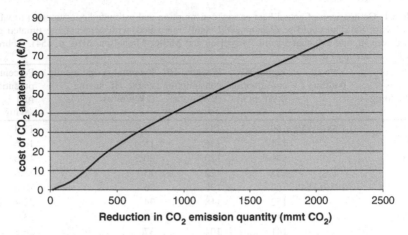

Fig. 6.4 A typical marginal cost curve for CO_2 abatement in Europe (Data from Morris et al. 2008, p. 17, converted to € from US$ at 0.77322 \$/€). For small reductions in the quantity of CO_2 emissions, the costs are low because it is assumed the 'easy cases' are tackled first. To reduce emissions by greater and greater amounts, more and more difficult cases need to be tackled, hence the cost per tonne of abatement rises

There is a vast literature on empirical studies of actual retrofit cases and their CO_2 abatement costs, so researchers have detailed knowledge of costs for a large range of energy-efficiency projects (Kossy and Ambrosi 2010; Sinn 2008). Costs range from a few euros to several hundred euros per tonne, depending on the type of facility being retrofitted and the depth of the retrofit.

Comparisons of different types of project have been made, and 'marginal cost abatement curves' are regularly produced (e.g. Morris et al. 2008). These show that a certain amount of CO_2 can be saved very cheaply, but that the more we attempt to save, the more expensive this becomes per tonne. For example, we can clean up the dirtiest East European factories for €10/t (Sinn 2008), but there are only a certain number of such factories. Once these are all CO_2-efficients, we have to move on to more expensive options. Marginal cost abatement curves for Europe and the USA currently only go up to about €60/t. At this moment, there is abundant CO_2-saving potential in projects cheaper than that: it would be economically inefficient for governments to invest in projects that cost more than €60/t. Figure 6.4 shows a typical CO_2 marginal cost abatement curve for Europe, from data offered by the MIT Joint Program on the Science and Policy of Global Change (Morris et al. 2008).

It is instructive to work out CO_2 abatement costs in thermal retrofit projects that fulfill German EnEV standards. When a retrofit is just economically viable, then (by definition) the cost of fuel saved is the same as the fuel price, currently around €0.10/kWh. Assuming a CO_2 intensity of 0.2 kg of CO_2 per kWh (as for natural gas), this equates to a CO_2 abatement cost of €500/t. Enseling and Hinz (2006), for example, report the retrofitting of a suite of apartment blocks in Ludwigshafen-am-Rhein, carried out in 2002–2003, when the legal standard for comprehensive retrofits was 150 kWh/m^{2}a. The additional thermal costs for apartments designed to reach this standard (in the event they actually reached 192 kWh/m^{2}a) were €0.0299/kWh, which equate to a CO$_2$ abatement cost of €149/t. The CO$_2$ abatement costs rose as the thermal standard increased. For apartments retrofitted to 30 kWh/m^{2}a the costs were €0.0906/kWh, a CO$_2$ abatement cost of €453/t. While thermal retrofits can bring valuable savings and improved comfort for households, it needs to be recognized that these costs can be much higher than the most expensive CO$_2$ abatement projects that would be considered worthwhile in industry and commerce. Hence, although a thermal retrofit may be 'economically viable', this does not necessarily imply that it is economically *efficient* in achieving the stated policy aim of reducing CO$_2$ emissions.

Similarly, Beecken and Schulze (2011) report the marginal costs of building an apartment block to a thermal standard 20% more energy efficient than a base standard of 60 kWh/m^{2}a. The increase in additional thermal costs that was required to improve the standard by this amount was €0.1931/kWh, equivalent to a CO$_2$ abatement cost of €965/t. Admittedly this was a higher standard than the 'economically viable' standard demanded by the EnEV, but as it qualified for Federal subsidies through the KfW bank, the CO$_2$ abatement at this level was funded by public money.

Kumbaroğlu and Madlener (2012) develop an interesting cost-benefit model to improve the economic efficiency of thermal upgrades. They suggest calculating costs and savings for the current fuel price and upgrade costs, then re-calculating for the expected fuel price and upgrade costs in several years time, then recalculating for a few years later, and so on. It could be that a retrofit in, say, 8 years time will bring a greater return on investment than one carried out today if the fuel price continues to rise. Some economically motivated households might see waiting as an economically efficient strategy.

Çomaklı and Yüksel (2003) propose an alternative to aiming for 'economically viable' thermal retrofits: aim instead for the 'economic optimum'. In a case study in the coldest cities in Turkey, they measure the costs and energy savings of various thicknesses of Styropor external wall insulation. For the city of Erzurum, which has twice as many heating degree-days (HDDs) as Berlin, the economically optimum thickness was just under 10 cm. Their extended model suggests that for a city with a climate like Berlin's (2,524 HDDs) this would be closer to 7 cm when the fuel price is around €0.10/kWh. Three years later Bolattürk (2006) found similar results, in a wider range of Turkish cities. For the city of Mardin, which has a similar climate to Berlin, the economic optimum was achieved with a thickness of 8 cm when the fuel price was around €0.10.

These figures cannot be transferred directly to German economic conditions. But the robust methodology that lies behind them leads us to suggest that Germany would benefit from similar studies. The economic optimum—where the highest CO$_2$ reduction is made for the lowest costs—is likely to be somewhat below the most energy-efficient economically viable thermal standard.

6.7 Economic Viability as a Policy Condition: The UK Green Deal

In Germany, the claim of economic viability—the promise of a payback—is used as a policy lever to motivate homeowners to undertake thermal retrofits. Whether it succeeds in doing this depends very much on how credible it appears to the homeowners. Evidence from interviews with German homeowners suggests that many who retrofit to high standards do so for reasons quite different from monetary payback, typically: to be more comfortable in the home; to reduce CO_2 emissions; and to enhance the quality of the home, much like when people renovate a kitchen or bathroom (Galvin 2011). Other interviewees say they have decided not to retrofit because they disbelieve the claims of economic viability.

But at a more strictly legal level, economic viability in Germany is a *condition*, designed to limit the powers of the building regulations: the condition of economic viability of a thermal standard has to be fulfilled, in order for that standard to be made compulsory for people doing renovations on their homes.

There is an interesting parallel and contrast here with the proposed Green Deal policy instrument in the UK. Here, the issue is not of the government's right to compel homeowners to do something, but of the homeowner's right to receive funding. If the condition of economic viability is fulfilled for certain retrofit was measures in a particular home, the homeowner will be able to apply for a loan to undertake these measures. The loan will be paid back in instalments along with fuel bill payments so that, assuming the measures are economically viable, the instalments will be no higher than the reduction in billing due to the fuel savings (Smith 2011). The so-called 'golden rule' of the Green Deal is that the post-retrofit monthly energy bill (including loan repayments and fuel costs) must be no higher than the pre-retrofit energy bill.

A further feature is that the loan will be drawn against the property, not its owner. This will avoid the barrier of low-income households not qualifying to take on debt. It will also free landlords from the dilemma of paying for thermal upgrades while the tenants get the benefit through reduced fuel bills, as the tenants' fuel bill will include the loan repayments. When a property is sold, the debt goes with the property.

Similar to the German situation, the Green Deal is intended to further the UK's climate goal of reducing GHG emissions by 80% by 2050 (DECC 2012a). How effectively it will do this will depend crucially on what level of thermal renovation is 'economically viable'. Some of the issues of economic viability, discussed above, will be similar for the UK, others different.

First, the prebound effect should be less of an issue in the UK, as methodologies are being devised to take account of actual household consumption in calculations of expected fuel savings. Nevertheless, these will not be based directly on the measured consumption of the homes applying for loans, but on a formula that modifies a dwelling's calculated energy rating according to the number of occupants and their income bracket. Developing a tool to assess occupant behavior was proved to be a very challenging task and delayed the implementation of the Green Deal.

Second, there will need to be clarity as to whether only the 'additional thermal costs' are to be financed by the loan, or whether the 'anyway costs' are to be included. This should not be an issue for cavity wall, ground floor and loft insulation, as practically all the expenses in these are additional thermal costs. For example, Green Deal literature estimates that for an average three-bedroom semi-detached house a loft insulation top-up would cost around £250 and bring annual savings of £45 (Smith 2011), easily fitting the criterion of economic viability. For windows, however, the total costs far exceed the additional thermal costs, so that fuel savings through window replacement often pay back as little as one-tenth the total cost of the measures (see Sect. 6.2). A similar situation arises with external wall insulation. One-third of English dwellings have noninsulated solid walls (ESDS 2011). If external wall insulation could be applied for £10,000, then even with an optimistic saving of £400 per year (Smith 2011), this would fail to conform to the rule of a 25-year payback period. Further, based on an average annual gas bill of £681 (DECC 2011), it is difficult to see how solid wall insulation would save £400 of this, or 59%. Compared to German EnEV, the Green Deal is a completely voluntary policy instrument and its effectiveness is going to depend on the attractiveness of the financial incentives it can offer, such as the interest rate.

Third, the timing of the fuel savings (see Sect. 6.4) will be an issue. If the loan repayments are on a table mortgage, the repayment amount will be the same each month for the 25-year period. But if fuel savings are calculated using the expected average cost of fuel over the next 25 years, the average value is not reached until the 14th year (or later if the assumed annual fuel price rise is high). For an annual fuel price rise of 4%, savings in the first year are only half the average value (see Fig. 6.1); for higher annual fuel price rises this is even lower. Hence, for the golden rule to work in the initial years, retrofit measures will fulfill this rule only if the *current* fuel price is used in the calculation of economic viability *and* the total monthly repayment, of interest plus principal, is used. This will reduce the depth of retrofit considerably.

However, this might have a positive effect on a fourth issue, the cost of CO_2 abatement. As we saw in Sect. 6.6, a lower depth of thermal retrofit often brings a higher ratio of fuel savings (and therefore CO_2 abatement) to money invested. Because of the golden rule—the constraint of ensuring that monthly investment loan repayments do not exceed monthly savings from reduced fuel consumption—there is a strong incentive to plan retrofit measures that get the maximum fuel saving for the minimum investment. This could bring considerable gains in economic efficiency.

The UK has 26.7 million homes (DECC 2012b). Approximately, 67% of these (17.9 million dwellings) were built before the 1970s energy crisis and the introduction of thermal regulations (ESDS 2011). Over 16 million of these have lofts, of which 1.9 million have <125 mm of loft insulation, while 15.8 million have cavity walls, of which 4.6 million are unfilled (DECC 2012b). This offers some 6.5 million insulation measures which would most likely be economically viable—perhaps not always to the stringent standards of the German EnEV, but in the more modest sense that would fit the golden rule criteria. Further, this would not necessarily lead to 'lock-in', where buildings are stuck with lower quality thermal performance for the next 40 years. If these low or modest level retrofit measures pay back within a decade, there is no economic loss in re-retrofitting, after than decade, to a higher standard. In the meantime CO_2 emissions have been reduced, fuel has been saved, and a household has possibly been warmer in winter.

6.8 Conclusions and Implications

German law seeks to protect building owners by limiting the thermal upgrade measures they are required to undertake, to those which are economically viable. But the claim of economic viability—the promise of a payback—is used as a policy lever to motivate homeowners to undertake thermal retrofits. Whether it succeeds in doing this depends very much on how credible it appears to the homeowners.

The way economic viability is framed in German policy brings four issues that must be faced by homeowners who are considering retrofitting: the economic viability formula only takes account of the 'additional thermal costs' and not the total costs of a retrofit; it does not allow for any extra construction costs required to change the form of the building, so that it can accommodate the legally required thickness of insulation; it produces a return on investment with increasing losses until after halfway through the lifetime of the upgrade measures and only pays back fully in the last month of this period; and it does not take account of the actual preretrofit consumption, which is likely to be significantly lower than the theoretical figure used in the calculations.

We propose that German policymakers revisit the rationale for setting such tight compulsory thermal standards for retrofits. In devising the EnEV, policymakers have used the economic viability criterion to fulfill other policy aims, such as the EU's '20-20-20' goal (see Chap. 3) and the 80% CO_2 reduction goal (see Chap. 2). This has proven problematic, however, for three main reasons: buildings are far more physically diverse than the official calculations of economic viability assume (see Chap. 4); occupants in non-retrofitted homes are generally far more fuel-thrifty than these calculations assume (see Chap. 5); and thermal retrofits to EnEV standards can be an economically inefficient way to reduce CO_2 emissions (Sect. 6.6).

This is not to suggest there is no place for thermal retrofits that reach and go beyond EnEV standards. But promoting them on the basis that they are 'economically viable' is problematic. There may be other good reasons for homeowners who have money to invest in their properties to retrofit to these standards, such as to be more comfortable in the home; to reduce CO_2 emissions regardless of personal cost; and to enhance the quality of the home. But in relation to a majority of homeowners, thermal retrofit policy would be better served by an emphasis on the 'economic optimum' than on that which is just 'economically viable'. This would require more flexible thermal retrofit regulations, and a different emphasis in training for home energy advisors.

References

Beecken C, Schulze S (2011) Energieeffizienz von Wohngebäuden: Energieverbräuche und Investitionskosten energetischer Gebäudestandards. Bauphysik 33(6):338–344

Bolattürk A (2006) Determination of optimum insulation thickness for building walls with respect to various fuels and climate zones in Turkey. Appl Therm Eng 26:1301–1309

Bundesregierung (2012) Energie für Deutschland: saubere, sichere, bezahlbare Energie: Deutschland verändert sich. Publikationsversand der Bundesregierung, Rostock

BMVBS (Bundesministerium für Verkehr, Bau und Stadtentwicklung) (2012) Interview mit Minister Ramsauer: "Wohnen muss bezahlbar bleiben", 4 July 2012. http://www.bmvbs.de/SharedDocs/DE/RedenUndInterviews/2012/BauenUndWohnen/bundesminister-dr-peter-ramsauer-im-interview-mit-der-augsburger-allgemeinen-am-04-07-12.html. Accessed 24 Sept 2012

Clausnitzer KD, Gabriel J, Diefenbach N, Loga T, Wosniok W (2009) Effekte des CO$_2$-Gebäudesanierungsprogramms 2008. Bremer Energie Institut, Bremen

Clausnitzer KD, Fette M, Gabriel J, Diefenbach N, Loga T, Wosniok W (2010) Effekte der Förderfälle des Jahres 2009 des CO$_2$-Gebäudesanierungsprogramms und des Programms „Energieeffizient Sanieren". Bremer Energie Institut, Bremen

Çomaklı K, Yüksel B (2003) Optimum insulation thickness of external walls for energy saving. Appl Therm Eng 23:473–479

DECC (Department of Energy and Climate Change) (2011) UK greenhouse gas emission statistics. http://www.decc.gov.uk/en/content/cms/statistics/climate_stats/gg_emissions/gg_emissions.aspx. Accessed 15 Dec 2011

DECC (Department of Energy and Climate Change) (2012a) Carbon emissions reduction target (CERT): paving the way for the green deal. http://www.decc.gov.uk/en/content/cms/funding/funding_ops/cert/cert.aspx. Accessed 27 Sept 2012

DECC (Department of Energy and Climate Change) (2012b) Estimates of home insulation levels in Great Britain, January 2012. http://www.decc.gov.uk/assets/decc/11/stats/energy/energy-efficiency/4537-statistical-releaseestimates-of-home-insulation-.pdf. Accessed 27 Sept 2012

DENA (Deutsche Energie-Agentur) (2012a) DENA-Studie: Energiesparendes Sanieren von Einfamilienhäusern rechnet sich. http://www.dena.de/presse-medien/pressemitteilungen/dena-studie-energiesparendes-sanieren-von-einfamilienhaeusern-rechnet-sich.html. Accessed 20 Oct 2012

DENA (Deutsche Energie-Agentur) (2012b) Energetische Sanierung ist wirtschaftlich. http://www.dena.de/aktuelles/alle-meldungen/energetische-sanierung-ist-wirtschaftlich.html Accessed 20 Oct 2012

DENA (Deutsche Energie-Agentur) (2012c) Energieeffiziente Gebäude. http://www.dena.de/themen/energieeffiziente-gebaeude.html. Accessed 26 Sept 2012

Diefenbach N, Loga T, Gabriel J, Fette M (2011) Monitoring der KfW-Programme „Energieeffizient Sanieren" 2010 und „Ökologisch/Energieeffizient Bauen" 2006–2010. Bremer Energie Institut, Bremen

EnEV (Energieeinsparverordnung) (2009) EnEV 2009—Energieeinsparverordnung für Gebäude. http://www.enev-online.org/enev_2009_volltext/index.htm. Accessed 26 Sept 2012

ESDS (Economic and Social Data Service) (2011) English house condition survey list of datasets. http://www.esds.ac.uk/findingData/ehcsTitles.asp. Accessed 12 Dec 2011

Enseling A, Hinz E (2006) Energetische Gebäudesanierung und Wirtschaftlichkeit—Eine Untersuchung am Beispiel des 'Brunckviertels' in Ludwigshafen. Institut Wohnen und Umwelt GmbH, Darmstadt

Feist W (1998) Wirtschaftlichkeitsuntersuchung ausgewählter Energiesparmaßnahmen im Gebäudebestand, Fachinformation PHI-1998/3. Passivhaus Institut, Darmstadt

Feist W (2009) Perspectives for the future—passive house technology. In: UNECE conference towards an action plan for energy efficient housing in the UNECE region, Vienna, 23–25 November 2009

Galvin R (2011) Discourse and materiality in environmental policy: the case of german federal policy on thermal renovation of existing homes. PhD Thesis, University of East Anglia. http://justsolutions.eu/Resources/PhDGalvinFinal.pdf. Accessed 19 Dec 2012

Galvin R (2013) Impediments to energy-efficient ventilation in German dwellings: a case study in Aachen. Energy Buildings 56:32–40. doi:10.1016/j.enbuild.2012.10.020

Hacker U (2009) Thesenpapier: Nutzverhalten im Mietwohnbereich. Institut Wohnen und Umwelt, Darmstadt

Hauser G, Maas A, Erhorn H, de Boer J, Oschatz B, Schiller H (2012) Untersuchung zur weiteren Verschärfung der energetischen Anforderungen an Gebäude mit der EnEV 2012—Anforderungsmethodik, Regelwerk und Wirtschaftlichkeit: BMVBS-Online-Publikation, Nr. 05/2012. BMVBS, Berlin

Kah O, Feist W, Pfluger R, Schnieders J, Kaufmann B, Schulz T, Bastian Z, Vilz A (2008) Bewertung energetischer Anforderungen im Lichte steigender Energiepreise für die EnEV und die KfW-Förderung, BBR-Online-Publikation 18/08. BMVBS/BBR, Herausgeber. http://www.bbr.bund.de/cln_015/nn_112742/BBSR/DE/Veroeffentlichungen/BBSROnline/2008/ON182008.html. Accessed 10 Nov 2011

Kossy A, Ambrosi P (2010) State and trends of the carbon market 2010. World Bank, Environment Department, Washington

Kumbaroğlu G, Madlener R (2012) Evaluation of economically optimal retrofit investment options for energy savings in buildings. Energy Buildings 49:327–334

Martinaitis V, Kazakevičiusb E, Vitkauskasb A (2007) A two-factor method for appraising building renovation and energy efficiency improvement projects. Energy Policy 35:192–201

Martinaitis V, Rogoža A, Bikmanienė I (2004) Criterion to evaluate the "Twofold Benefit" of the renovation of buildings and their elements. Energy Buildings 36(1):3–8

Michelsen C, Müller-Michelsen S (2010) Energieeffizienz im Altbau: Werden die Sanierungspotenziale überschätzt? Ergebnisse auf Grundlage des ista-IWH-Energieeffizienzindex. Wirtschaft im Wandel, 9, 447-455. http://www.iwh-halle.de/d/publik/wiwa/9-10-5.pdf. Accessed 28 Nov 2011

Morris J, Paltsev S, Reilly J (2008) Marginal abatement costs and marginal welfare costs for Greenhouse gas emissions reductions: results from the EPPA model. Massachusetts Institute of Technology, Cambridge

Simons H (2012) Energetische Sanierung von Ein- und Zweifamilienhäusern Energetischer Zustand Sanierungsfortschritte und politische Instrumente. Bericht im Auftrag des Verbandes der Privaten Bausparkassen e.V. Empirica, Berlin

Sinn HW (2008) Das Grüne Paradoxon: Plädoyer für eine illusionsfreie Klimapolitik. Verlag, Berlin

Smith L (2011) The green deal: standard note SN/SC/5763. House of commons library, science and environment section, London. http://www.parliament.uk/briefing-papers/SN05763.pdf. Accessed 15 Dec 2011

Sunikka-Blank M, Galvin R (2012) Introducing the prebound effect: the gap between performance and actual energy consumption. Building Res Inf 40:260–273

Tuominen P, Klobut K, Tolman A, Adjei A, de Best-Waldhober M (2012) Energy savings potential in buildings and overcoming market barriers in member states of the European Union. Energy Buildings 51:48–55

Chapter 7
Why is Domestic Heating Fuel Consumption Falling in Germany?

Abstract Heating energy consumption has been falling steadily in Germany since 2000. The most recent reliable figures show it fell from 669 TWh to 550 TWh in the years 2000–2009, a reduction of 119 TWh, or 18%. A number of different factors could be contributing to this: replacement of old dwellings with energy-efficient new builds; thermal retrofits of existing homes; and non-technical factors such as household behavior and demographics. There is now considerable data on numbers of new builds and retrofits and their heating energy consumption, but until now there has been no attempt to analyze this data to disaggregate the contributions of each of these factors to the steady fall in consumption. This chapter attempts such an analysis. We find that there was no net reduction from replacement of old with new dwellings, as the latter outnumber the former by about 2 to 1. Thermal retrofits appear to account for about 14 TWh, or 12% of the reduction. This leaves a residual fall of 105 TWh, or 88% of the total, which appears to be due to non-technical factors such as behavior change and demographics, representing reduced annual expenditure on heating fuel of around €7.35 billion and reduced annual CO_2 emissions of 21 million tons. This points to the need for more research, first to confirm these results, and second to investigate what motivates many nonretrofitting householders to save heating fuel.

Keywords Fuel savings · CO_2 emission reduction · Thermal retrofit · Household behavior · Price elasticity

7.1 Introduction

Figures released by the Federal Statistics Office (Destatis) reveal that energy consumption for domestic heating fell from 669 TWh in the year 2000 to 550 TWh in 2009 (Destatis 2010a). These figures are 'temperature adjusted', and hence free, at least theoretically, from the influence of variations in winter weather from year to year. This represents a fall in domestic heating consumption of 18%

R. Galvin and M. Sunikka-Blank, *A Critical Appraisal of Germany's Thermal Retrofit Policy*, Green Energy and Technology, DOI: 10.1007/978-1-4471-5367-2_7, © Springer-Verlag London 2013

in 9 years, which is also reflected in other statistical surveys. For example, CO_2Online collects data on actual energy use from 950,000 households through its Web portal. It finds their rate of heating fuel consumption fell steadily over the years 2000–2009 from 164 kWh/m^{2}a to 135 kWh/m^{2}a, also a fall of 18% (CO_2Online 2010). As this is a self-selected sample it is not as reliable as the wider samples used by Destatis, but the trend is comparable.

The Destatis figures show a reduction in national heating consumption of 119 TWh, indicating that German householders spent €8 billion less on heating fuel in 2009 than they would have if their consumption had stayed steady at the 2000 level (the price of heating fuel in 2009 was approximately €0.07/kWh). They emitted 30 million tons less CO_2 emissions than they would have at 2000 levels, even though the number of occupied dwellings increased by 3.3% in 2000–2009.

It is also interesting that the water-heating component of this consumption stayed steady in 2000–2009 at 82 TWh per year (Destatis 2010a). Hence space-heating consumption fell by the larger factor of just over 20%. However, in keeping with the approach of this book, we will consider space and water heating together here.

It is often assumed in policy literature that heating fuel reductions come entirely from technical upgrade measures: thermal retrofits of existing homes and replacement of older homes with energy-efficient new builds (e.g. BMVBS 2012; BMWi 2010). The national statisticians who published the Destatis figures, however, made the following comment:

> On the one hand this (reduction) can be attributed to improved insulation and heating technology. But on the other hand, it is also the result of household savings as a reaction to significant rises in the cost of heating energy (Destatis 2010a, authors' translation).

However, so far there has been no attempt to quantify the proportions of domestic heating fuel savings being made from behavioral as compared to technical measures. Rehdanz (2007) noted that it would be useful to disaggregate these proportions, and Gram-Hanssen (2010, 2011) argues from empirical evidence that both behavioral and technical measures are significant in attempts to lower household energy consumption. The Federal Environment Office (UBA 2006) surveyed households in 1995–2005 and concluded that changes in user behavior were making a significantly greater contribution to energy savings than technical, energy-efficiency improvements. There is now sufficient empirical evidence for us to be able to estimate fuel savings made from new builds and retrofits to a reasonably accurate degree, so that we can see what savings are not accounted for by these technical measures. This is what we set out to do in this chapter.

Our methodology is set out schematically in Fig. 7.1. The quantities denoted by variables *A*, *B*, *C*, etc., which are displayed on this schematic, are given in Table 7.1. Other quantities we will be referring to in our analysis are given in Table 7.2.

Starting at the top of the schematic, in Sect. 7.2 we estimate the number of dwellings newly constructed in 2000–2009 and their average heating consumption in 2009, to give the quantity F. Then in Sect. 7.3 we estimate the total number of

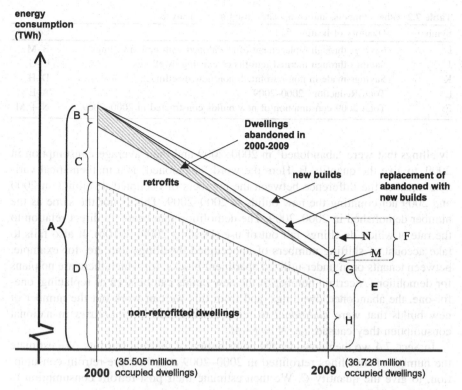

Fig. 7.1 Schematic showing methodology for disaggregating the contribution of four classes of dwelling to national totals of heating energy consumption: homes existing in 2000 that were not retrofitted in 2000–2009; homes that were retrofitted during this period; homes abandoned during this period; and homes newly built during this period

Table 7.1 Quantities of heating energy displayed in Fig. 7.1

Symbol	Quantities of heating fuel consumed in 2000 by	Quantities of heating fuel consumed in 2009 by
A	All dwellings	
B	Dwellings abandoned in 2000–2009	
C	Dwellings upgraded in 2000–2009	
D	Dwellings not upgraded in 2000–2009	
E		All dwellings
F		Dwellings constructed in 2000–2009
G		Dwellings upgraded in 2000–2009
H		Dwellings not upgraded in 2000–2009
M		New builds replacing abandoned dwellings
N		New builds additional to replacements

Table 7.2 Other symbols, and composites, used in the analysis

Symbol	Quantity of heating fuel	Equal to
I	Savings through replacement of abandoned with new dwellings	B–M
J	Savings through thermal retrofits of existing dwellings	C–G
K	Savings made in non-retrofitted, non-new dwellings	D–H
L	Total Reductions 2000–2009	A–E
F	Total 2009 consumption of new builds constructed in 2000–2009	N + M

dwellings that were 'abandoned' in 2000–2009 and their average consumption in 2000, to give the quantity B. Here the word 'abandoned' is a mathematical variable equal to the difference between the numbers of occupied dwellings in 2000 and 2009 not counting the new builds in 2000–2009. This is not the same as the number demolished in 2000–2009. The demolition rate bears no direct relation to the rate at which dwellings fall out of use (Destatis 2004, 2010b). It also fails to take account of shifting numbers of unoccupied dwellings that are, for example, between tenants or in under-utilized apartment blocks, for which there are no plans for demolition. A certain number of the new builds can be seen as replacing, one-for-one, the abandoned dwellings, and from this we can work out the number of new builds that were *additional* to replacements, and the *increases* in national consumption they caused, i.e. $N = F - M$.

In Sect. 7.4 we then consider the contribution of thermal retrofits. We estimate the number of dwellings retrofitted in 2000–2009 and their pre-retrofit consumption, to give the quantity C. We then estimate their post-retrofit consumption to give the total G.

In Sect. 7.5 we bring all these figures together to estimate the reductions in consumption in 2000–2009 that were not due to retrofits or new builds, namely:

$$K = D - H = (A - B - C) - (E - F - G)$$

We draw conclusions from the results of this analysis and make recommendations for policymakers in Sect. 7.6.

It should be noted that we are looking at a 9-year time span in this analysis. Our figures refer to the period 1 July 2000 to 30 June 2009, or the 9-year period closest to these dates which each dataset used in the analysis relates to. We choose 2009 as our cut-off year because this is the most recent year for which figures are available. Domestic heating consumption in Germany peaked in 2000, so this seems a suitable starting date.

7.2 New Builds in 2000–2009

Destatis (2010a, b) figures reveal that 2,622,327 new dwellings were completed in 9 years from January 2000 to December 2008. These had an average 'useful' area of 114 m^2 (Destatis 2010b), giving a total new 'useful' area of 299,000,000 m^2 (see Chap. 5, Box 5.1 for the difference between living area and useful area).

We now estimate how much these dwellings were consuming in 2009. Prior to October 2012, the average[1] maximum permissible heating fuel consumption for new builds was 145 kWh/m^2a; from October 2002 to September 2009 it was 100 kWh/m^2a; and thereafter 70 kWh/m^2a. Federal subsidies led to many of these homes being designed for higher thermal standards (DENA 2012), which would have lowered the average consumption in each of these periods.

The most comprehensive study of the energy performance of new homes built in this period is by Greller et al. (2010). These authors analyzed the metered consumption of 110,000 gas and oil-heated homes built from 1977 to 2006, including 25,650 in the period 2000–2006. They found an average heating fuel consumption of 95 kWh/m^2a for those built in the years 2000–2006 inclusive (Greller et al. 2010) and a falling trend from 100 kWh/m^2a to 90 kWh/m^2a over this period. Assuming this trend continued at the same rate for the next 2 years, the average for the 9-year period would be 93 kWh/m^2a. Hence we would estimate the heating energy consumption of these buildings, in 2009, at $2,622,327 \times 93$ $kWh/m^2a \times 114$ $m^2 = 27.8$ TWh.

The three other main sources of heating fuel—district heating, wood, and electricity—make up 7, 7.5 and 3% of heating consumption, respectively (Schloman et al. 2004). District heating is associated with consumption 10% lower than average, and wood and electricity tend to consume 10% higher. Hence the actual total might be slightly higher than the 27.8 TWh estimated above. We suggest a figure of 29 TWh is indicated, so that our variable $F = 29.0$ TWh.

7.3 Dwellings Abandoned in 2000–2009

The term 'dwellings abandoned in 2000–2009' is a mathematical variable, equal to the sum of the dwellings occupied in 2000 (35,505,000) and the dwellings newly built in 2000–2009 (2,622,327), less the dwellings occupied in 2009 (36,728,000). This puts the net number of dwellings abandoned during this period at 1,399,000 (to the nearest 1000).

There are no existing studies on the heating consumption of dwellings that become unoccupied, but there are good reasons to assume their consumption was higher than the national average. Dwellings become unoccupied for two main reasons in Germany: internal migration; and poor quality of buildings. Migration is most likely the major factor, as there have been large internal population shifts in Germany over the last 20 years, mostly from East to West and North to South. The housing stock in former East Germany was mostly of pre-World War II thermal quality prior to reunification, and the Communist era *Plattenbau* (prefabricated slab) apartment blocks were of low thermal quality (Flockton 1998). In the western

---

[1] As we noted in Chap. 2, the Energy Saving Regulations (EnEV) prescribe a range of maximum consumption figures for buildings depending on their geometry, size and connection to other buildings.

states, emigration occurs mostly from industrial areas, such as the Ruhr Valley. In regions of falling population such as these, relatively few homes have been built in recent decades. Greller et al. (2010) show that the average consumption of pre-World War II dwellings is around 165 kWh/m²a and that of dwellings built in 1946–1964 around 160 kWh/m²a. Schröder et al. (2011), using similar methodology and datasets, estimate total national average consumption at 149 kWh/m²a. This would put the consumption of pre-World War II dwellings about 10% higher than the national average, and as most abandoned dwellings are of older stock, it would seem reasonable to assume their average consumption was around 10% above the national average.

As we noted above, the national domestic heating energy consumption in 2000 was 669 TWh and there were 35,505,000 occupied dwellings, so that average consumption was 18,842 kWh/a per dwelling. If the buildings that were later abandoned were consuming 10% above the average, their consumption would have been 20,727 kWh/a per dwelling. As there were 1,399,000 of these dwellings, their total heating consumption in 2000 would have been 29.0 TWh. Hence we set $B = 29.0$ TWh.

We note, also, that 1,399,000 of the new builds in 2000–2009 can be seen as replacing abandoned dwellings. If the replacements' useful area was 114 m² with heating consumption 93 kWh/m²a (see above), this group's total consumption was 14.8 TWh. Hence we set $M = 14.8$ TWh.

Therefore the heating consumption of new builds that were not replacing abandoned dwellings was the difference between the total new build consumption of 29.0 TWh (see Sect. 7.2) and this figure, i.e. $F-M = 29.0-14.8 = 14.2$ TWh. Hence we set $N = 14.2$ TWh.

An interesting result here is that the fall in consumption from dwellings being abandoned, at 29 MWh, is the same as the rise in consumption from new builds. The net effect of these two sectors on fuel savings is to cancel each other out, as there were approximately twice as many new dwellings as abandoned dwellings.

7.4 Dwellings Retrofitted in 2000–2009

7.4.1 General Considerations

The most difficult part of our analysis is to assess the fuel consumption reductions due to retrofits. Not only are the numbers of retrofitted dwellings in dispute, but there is a variety of degrees of retrofit, up to a full project including wall, roof and basement-ceiling insulation; window and door replacement; addition of ventilation system (with or without heat recovery); boiler and radiator replacement; boiler and heating system fine tuning; and installation of solar collectors. Further, as there is no inspection of thermal (or other) building standards, the authorities do not automatically have records of what retrofitting has taken place and we know from unofficial sources that a certain amount of 'sub-standard' retrofitting is carried out illegally, often in broad daylight.

We will consider national summary figures from peer-reviewed papers and from research institutes commissioned by the Federal government. Since the precise numbers are in dispute in this section, we will estimate the range within which the true value is highly likely to fall—the 95% confidence interval—and carry these results forward to our final totals. Further, we will at times speak of 'equivalent floor area' or 'equivalent dwellings' rather than 'number of dwellings', as some studies estimate the savings due to partial retrofits in terms of their equivalent number of dwellings or floor area fully retrofitted.

7.4.2 Annual Rate of Thermal Retrofits

The German Energy Agency (DENA) estimates the annual rate of thermal retrofits at 0.8% of the residential building stock per year over the past decade (Stolte 2011), i.e. 7.2% of the residential building stock over 2000–2009. Since the average number of dwellings (occupied and unoccupied) in 2000–2009 was 39.2 million, this is equivalent to 2.82 million dwellings.

Diefenbach et al. (2010) found a retrofit rate of 0.83% per year in a survey of 7,500 building owners, equivalent to 7.5% over 2000–2009, or 2.94 million dwellings if extrapolated nation-wide, though this is a limited sample size.

Friedrich et al. (2007) estimate the equivalent living area of existing homes retrofitted with insulation, new windows, or new boilers in each of the years 2000–2006 as amounting to an accumulated total of 5.9% of the residential stock over these 7 years, which would amount to 7.6% if it continued for the remainder of the 9 years, or 2.98 million dwellings.

Weiss et al. (2012) estimate an annual retrofit rate for walls at 0.8% and roofs at 1.2%. Weighting walls and roofs according to their average thermal impact[2] in the ratio 3:2 would give an annual retrofit rate of 0.96%, assuming windows and boilers were upgraded correspondingly. This would amount to 8.6% of the residential stock over 9 years, or 3.37 million dwellings.

While the figures from each of these studies do not precisely agree, they are sufficiently close to be used in our analysis. We note their average, of 3.03 million dwellings, with a standard deviation of 0.24 million, or 8% of the total. Taking them together gives a 95% confidence interval of 2.56–3.50 million dwellings, suggesting that it is highly unlikely that the number of equivalent comprehensively retrofitted dwellings fell outside this range.

---

[2] The term 'thermal impact' means the relative contribution of each feature to the thermal quality of the building, based on the area each contributes to the building envelope, and typical U-values of each.

7.4.3 Pre-retrofit Consumption in 2000

We now estimate the heating fuel savings achieved through these retrofits. To begin with, there is no reliable data as to their actual pre-retrofit heating fuel consumption. The German Energy Agency targets dwellings with a calculated consumption of 225 kWh/m^{2}a or more for retrofitting (Stolte 2011). Taking the prebound effect into account (see Chap. 5), this would give an actual pre-retrofit consumption of around 160 kWh/m^{2}a. Data from CO$_2$Online indicates an average pre-retrofit calculated consumption of 260 kWh/m^{2}a for detached houses and 200 kWh/m^{2}a for all other buildings, with actual measured pre-retrofit consumption 160 kWh/m^{2}a and 150 kWh/m^{2}a, respectively. However, as previously noted, the self-selection bias of CO$_2$Online's data might make these actual consumption figures lower than national averages for pre-retrofit dwellings.

Retrofits in Germany are mostly performed on homes built prior to the first thermal regulations, in 1977. As we noted in Sect. 7.3, the average consumption of these homes is most likely some 10% higher than national average consumption, i.e. around 164 kWh/m^{2}a. Further, as the government targets the least thermally efficient homes for thermal upgrading, it is reasonable to assume that pre-retrofit consumption in the cohort being retrofitted is somewhat higher than the national average of that group, say a further 10%.

Hence, it is likely that the pre-retrofit actual consumption of dwellings that were retrofitted in 2000–2009 was around 165–174 kWh/m^{2}a. We cautiously estimate the 95% confidence interval at this range.

Since the average useable area in 2000 was 110 m^2 (living area 84 m^2), this gives a range of 46.5–66.9 TWh TWh for the heating fuel consumption in 2000, of dwellings that were retrofitted in 2000–2009. Hence our variable $C = 46.5–66.9$ TWh.

7.4.4 Consumption Reduction per Retrofitted Dwelling

We now estimate the reduction in heating fuel consumption achieved through these thermal retrofits. Schröder et al. (2011) investigated the heating energy consumption of residential buildings of two or more dwellings throughout Germany for 2004–2008. Buildings retrofitted or constructed since 1995 showed an average consumption of 110 kWh/m^{2}a, while those constructed prior to 1995 and not subsequently retrofitted consumed an average of 145 kWh/m^{2}a—a difference of 35 kWh/m^{2}a, or 24%.

A further nation-wide empirical study (Schröder et al. 2010) gives cumulative frequency distribution curves for the heating energy consumption throughout Germany of retrofitted and nonretrofitted apartment blocks of floor area larger than 700 m^2. The mean is 140 kWh/m^{2}a for those completely nonretrofitted and 90 kWh/m^{2}a for those comprehensively thermally retrofitted, a fuel saving of 50 kWh/m^{2}a, or 36%.

Regarding smaller buildings, Walberg et al. (2011) investigated the nationwide retrofit performance of 5 classes of 1–2 dwelling buildings and found an average measured consumption reduction of 26%.

A comprehensive figure is offered by Tschimpke et al. (2011), of an average of 38% reductions achieved through retrofits of all building types, though this refers to calculated pre-and post-retrofit consumption figures. As we saw in Chap. 2, Clausnitzer et al. (2009, 2010) estimate the calculated saving in retrofits with KfW subsidies at an average of 33%. We estimated that, taking the prebound effect into account, this would indicate actual savings of around 25%. We noted that the German Energy Agency also estimates actual savings at 25%. Figures from CO_2Online indicate actual savings of around 33%, and again this could be higher than the national average due to self-selection bias in the data. However, we cannot ignore Schröder et al. (2010) figure of 36% for large apartment blocks. Hence, we would suggest a mean of around 30% savings for equivalent comprehensive retrofits carried out in 2000–2009, with a possible 95% confidence interval of 25–36%.

We now bring these percentages together with the pre-retrofit consumption range estimated above. The maximum reduction would arise if the maximum pre-retrofit consumption were reduced by the highest of these percentages, i.e. 66.9 TWh reduced by 36%, a reduction of $J_{max} = 24.1$ TWh. The minimum reduction would arise if the minimum pre-retrofit consumption were reduced by the lowest of these percentages, i.e. 46.5 TWh reduced by 25%, a reduction of $J_{min} = 11.6$ TWh. The mid-point between these is $J_{mid} = 17.9$ TWh. We use these values for $C - D$ in Equation 1.

7.5 Reductions Due to Nontechnical or Unexplained Factors

We now have sufficient information to estimate the savings in 2000–2009 that are not explained by thermal retrofits or new builds. These are represented by the variable K, where (recalling Equation 1):

$$K = D - H = (A - B - C) - (E - F - G)$$
$$= (A - B) - (E - F) - (C - G)$$
$$= (A - B) - (E - F) - J$$

Hence the lowest likely value of K (taking the highest value of J) is:

$$K = (669.0 - 29.0) - (550.0 - 29.0) - 24.1$$
$$K_L = 94.9 \text{ TWh}$$

And the highest likely value of K (taking the lowest value of J) is:

$$K = (669.0 - 29.0) - (550.0 - 29.0) - 11.6$$
$$K_H = 107.4\,\text{TWh}$$

This suggests that heating fuel reductions in 2000–2009 that are not attributable to retrofits or new builds are most likely to be in the range 94.9–107.4 TWh. This represents 79.7–90.3% of the total reductions, or a middle value of 85.0%.

These figures also give results for D and H, the consumption in 2000 and 2009, respectively of dwellings existing in 2000 that were not retrofitted in 2000–2009.

$$D = A - B - C$$
$$D_{\text{max}} = 669.0 - 29.0 - 46.5 = 593.5\,\text{TWh}$$
$$D_{\text{min}} = 669.0 - 29.0 - 66.9 = 573.1\,\text{TWh}$$
$$H = E - F - G = E - F - (C - J)$$
$$H_{\text{max}} = 550.0 - 29.0 - (46.5 - 11.6) = 486.1\,\text{TWh}$$
$$H_{\text{min}} = 550.0 - 29.0 - (66.9 - 24.1) = 478.2\,\text{TWh}$$

Hence the percentage fall in consumption 2000–2009 of nonretrofitted dwellings lies between 16.6% and 18.1%, with a mid-range value of 17.4%.

We note that this is just over half the percentage fall in consumption being achieved through thermal retrofits. Table 7.3 gives our estimates of the national totals for each of the main quantities we have considered in the analysis. Table 7.4 gives the percentage savings made by each sector, assuming mid-range estimates for retrofits.

Table 7.3 Estimates of quantities of heating energy for each variable in the analysis

Symbol	Estimate (TWh)	Quantities of heating fuel consumed in 2000 by	Quantities of heating fuel consumed in 2009 by
A	669.0	All dwellings	
B	29.0	Dwellings abandoned in 2000–2009	
C	46.5–66.9	Dwellings upgraded in 2000–2009	
D	573.1–593.5	Dwellings not upgraded in 2000–2009	
E	550.0		All dwellings
F	29.0		Dwellings constructed in 2000–2009
G	29.8–24.8		Dwellings upgraded in 2000–2009
H	478.2–486.1		Dwellings not upgraded in 2000–2009
M	14.8		New builds replacing abandoned dwellings
N	14.2		New builds additional to replacements

Table 7.4 Mid-range estimates of heating fuel savings in 2000–2009 in each sector of the housing stock

Symbol	Sector	Savings achieved (TWh)	Percentage share of total national savings	Savings as percentage of 2000 level for this sector
L	All dwellings	119	100	18
I	Replacement	0	0	0
J	Retrofits	18	15	30
K	Behavior	101	85	17.4

A peculiarity of the analysis is that the reduction in consumption due to abandonments is completely cancelled out by the additional consumptions from new builds. Although new builds are highly energy-efficient compared with abandoned dwellings, there are a lot more of them. If demographic and populations trends continue much as they have for the past decade, this is not likely to change significantly. The trend in Germany is toward fewer persons per household and larger floor area per dwelling (BMVBS 2011).

Nevertheless, some of the consumption reduction in existing, non-retrofitted homes could be due to these trends. Because new builds are increasing the total number of dwellings while the population is falling, some existing households are losing members. They have fewer occupants, so fewer rooms need to be heated, common rooms need fewer hours of heating, and less hot water is consumed. In a sense, existing homes are exporting some of their heating consumption to new builds.

The high estimate our analysis has produced, of the contribution of non-retrofitted dwellings to national consumption reductions, is surprising. However, even if average savings through retrofits were double the percentages we estimate, say 60%, this would still only reduce the non-retrofit contribution to 69% of the total fall. Even if savings per retrofit were double *and* the annual rate of retrofits were twice as high as our estimate for 2000–2009, then non-retrofit contribution would still be over 40%. Hence, even in these extreme scenarios, non-retrofitting households are making a mathematically significant contribution to heating consumption reductions.

7.6 Conclusions and Implications

Drawing on national statistics, peer-reviewed empirical studies, and government commissioned reports from research institutes, we estimated the contributions of new builds, abandonment of dwellings and thermal retrofits to the 18% fall in German domestic heating consumption from 669 TWh in 2000 to 550 TWh in 2009. We found that new builds added 29.0 TWh to the 2000 total but that this was cancelled out by consumption reductions due to dwelling abandonment. Thermal retrofits reduced national consumption by 11.6–24.1 TWh in the period

2000–2009, leaving the remaining reduction of 94.9–107.4 TWh to non-technical or unexplained factors. We suggested that, even if savings through retrofits were four times our highest estimate, this would still leave an unexplained residual fall of around 50 TWh, or 42% of the total fall.

Part of this residual fall can possibly be explained by demographic and lifestyle changes, but there appears to be no technical explanation for the bulk of the residual reductions. The most natural conclusion is that these reductions are largely caused by changes in user behavior. It is likely that a large proportion of householders are heating fewer rooms, or for shorter periods, or to lower temperatures, or some combination of these than they used to—or ventilating less adequately, a possibility explored in Galvin (2013).

We already know that there is a large potential in German homes for saving heating energy through behavior changes and we know the strategies occupants can use to achieve this (UBA 2006). What we do not know is why some households do this, and others less so, or not at all. One of the motivating factors is financial savings. Rehdanz (2007) has shown that domestic heating consumption in Germany is sensitive to fuel price rises. The price of heating fuel rose by 52.5% in 2000–2009 (BMWi 2011a, b), representing an average annual increase of 4.8% (since $1.048^9 = 1.525$). Using a formula derived by Galvin and Sunikka-Blank (2012), this represents a year-on-year fuel price elasticity for households in non-retrofitted dwellings of –0.425. This means that in each year, every 1% rise in fuel price is associated with an aggregate 0.425% fall in heating fuel consumption by these households, a topic further explored in Chap. 8).

This is not to claim that fuel price increases were the only factor motivating consumption reductions. Rising environmental awareness or other motivational factors, yet to be explored, might also play a role. Further, the price elasticity figure here is an average value. Some households might not have responded at all to fuel price rises; others might have responded much more sharply.

According to our estimates, nonretrofitting households consumed around 90 TWh less heating fuel in 2009 than in 2000, worth €7.35 billion in 2009 prices, and their heating systems emitted 21 million tonnes less CO_2. This was cost-free CO_2 abatement, compared with an average price of €250 per tonne of CO_2 saved through thermal retrofits to EnEV standard (see Chap. 6). We suggest that it would be profitable for the German government to take more notice of the significant fuel savings being made in non-retrofitted households.

References

BMVBS (Bundesministerium für Verkehr, Bau und Stadtentwicklung), (2011) Wohnen und Bauen in Zahlen 2010/2011: 6, Auflage, Stand: Mai 2011. BMVBS, Berlin
BMVBS (Bundesministerium für Verkehr, Bau und Stadtentwicklung) (2012) Energieeffiziente Gebäude und Städte. BMVBS, Berlin. http://www.bmvbs.de/DE/BauenUndWohnen/EnergieeffizienteGebaeude/energieeffiziente-gebaeude_node. Accessed 12 Nov 2012

BMWi (Bundesministerium für Wirtschaft und Technologie) (2010) Eckpunkte Energieeffizienz. http://www.bmwi.de/BMWi/Redaktion/PDF/E/eckpunkte-energieeffizienz,property=pdf, bereich=bmwi,sprache=de,rwb=true.pdf. Accessed 12 Nov 2012

BMWi (Bundesministerium für Wirtschaft und Technologie) (2011a) Energiedaten: ausgewählte Grafiken, Stand: 15.08.2011a. http://www.bmwi.de/BMWi/Navigation/Energie/Statistik-und-Prognosen/energiedaten.html. Accessed 14 Dec 2011

BMWi (Bundesministerium für Wirtschaft und Technologie) (2011b). Zahlen und Fakten: Energidaten: Nationale und Internationale Entwicklung, Bundesministerium für Wirtschaft und Technologie, Referat III C 3. http://www.bmwi.de/BMWi/Navigation/Energie/Statistik-und-Prognosen/energiedaten.html. Accessed 5 Dec 2010

Clausnitzer KD, Gabriel J, Diefenbach N, Loga T, Wosniok W (2009) Effekte des CO_2-Gebäudesanierungsprogramms 2008. Bremer Energie Institut, Bremen

Clausnitzer KD, Fette M, Gabriel J, Diefenbach N, Loga T, Wosniok W (2010) CO2Online (2010) Heizenergieverbrauch in Deutschland: eine quantitative Auswertung von 950,000 Gebäudedatensätzen erhoben von der co2Online gemeinnützige GmbH. www.co2online.de. Accessed 19 Dec 2012

DENA (Deutsche Energie-Agentur) (2012) Thema Enegie: Lüftungsarten: Nicht zu viel und nicht zu wenig—Lüftung mit Technik. http://www.thema-energie.de/heizung-heizen/lueften-kuehlen/lueftungsarten.html. Accessed 7 April 2012

Destatis (2004) Bautätigkeit und Wohnungen: Bestand an Wohnungen; Fachserie 5/Reihe 3, 2004. Statistisches Bundesamt, Wiesbaden

Destatis (2010a) Energieverbrauch der privaten Haushalte für Wohnen rückläufig: Pressemitteilung Nr.372 vom 18.10.2010. Statistisches Bundesamt, Wiesbaden. http://www.destatis.de/jetspeed/portal/cms/Sites/destatis/Internet/DE/Presse/pm/2010/10/PD10__372__85,templateId=renderPrint.psml. Accessed 28 Jan 2012

Destatis (2010b) Bauen und Wohnen: Baugenehmigungen/Baufertigstellungen Lange Reihen z. T. ab 1949; Pub. 2010b. Statistisches Bundesamt, Wiesbaden

Diefenbach N, Cischinsky H, Rodenfels M, Clausnitzer KD (2010) Datenbasis Gebäudebestand Datenerhebung zur energetischen Qualität und zu den Modernisierungstrends im deutschen Wohngebäudebestand. Institut Wohnen und Umwelt/Bremer Energie Institut, Darmstadt. http://www.iwu.de/fileadmin/user_upload/dateien/energie/klima_altbau/Endbericht_Datenbasis.pdf. Accessed 21 Jan 2012

Flockton C (1998) Housing situation and housing policy in East Germany. German Politics 7(3):69–82

Friedrich M, Becker D, Grondy A, Laskosky F, Erhorn H, Erhon-Kluttig, Hauser G, Sager C, Weber H (2007) CO_2-Gebäudereport 2007, im Auftrag des Bundesministeriums für Verkehr, Bau und Stadtentwicklung (BMVBS). Fraunhofer Institut für Bauphysik, Stuttgart

Galvin R (2013) Impediments to energy-efficient ventilation in German dwellings: a case study in Aachen. Energy Buildings 56:32–40. doi:10.1016/j.enbuild.2012.10.020

Galvin R, Sunikka-Blank M (2012) Including fuel price elasticity of demand in net present value and payback time calculations of thermal retrofits: case study of German dwellings. Energy Buildings 50:219–228

Gram-Hanssen K (2010) Residential heat comfort practices: understanding users. Building Res Inf 38(2):175–186

Gram-Hanssen K (2011) Households' energy use—which is the more important: efficient technologies or user practices? Paper presented at the world renewable energy congress, Linköping, 8–13 May 2011

Greller M, Schröder F, Hundt V, Mundry B, Papert O (2010) Universelle Energiekennzahlen für Deutschland—Teil 2: Verbrauchskennzahlentwicklung nach Baualtersklassen. Bauphysik 32:1–5

Rehdanz K (2007) Determinants of residential space heating expenditures in Germany. Energy Econ 29:167–182

Schloman B, Ziesling HJ, Herzog T, Broeske U, Kaltschnitt M, Geiger B (2004) Energieverbrauch der privaten Haushalte und des Sektors Gewerbe, Handel, Dienstleistungen (GHD), Projekt Nr

17/10, Abschlussbericht an das Ministerium für Wirtschaft und Arbeit. Fraunhofer Institut für Systemtechnik und Innovationsforschung, Karlsruhe. http://isi.fraunhofer.de/isi-de/e/projekte/122s.php. Accessed 26 Nov 2011

Schröder F, Engler HJ, Boegelein T, Ohlwärter C (2010) Spezifischer Heizenergieverbrauch und Temperaturverteilungen in Mehrfamilienhäusern—Rückwirkung des Sanierungsstandes auf den Heizenergieverbrauch. HLH 61(11):22–25. http://www.brunata-metrona.de/fileadmin/Downloads/Muenchen/HLH_11-2010.pdf. Accessed 8 Dec 2011

Schröder F, Altendorf L, Greller M, Boegelein T (2011) Universelle Energiekennzahlen für Deutschland: Teil 4: Spezifischer Heizenergieverbrauch kleiner Wohnhäuser und Verbrauchshochrechnung für den Gesamtwohnungsbestand. Bauphyisk 33(4):243–253

Stolte C (2011) The German experience of thermal renovation. Presentation at the conference cutting carbon costs: our big energy battle, London School of Economics, London, 8 Nov 2011

Tschimpke O, Seefeldt F, Thamling N, Kemmler A, Claasen T, Gassner H, Neusüss P, Lind E (2011) Anforderungen an einen Sanierungsfahrplan: Auf dem Weg zu einem klimaneutralen Gebäudebestand bis 2050. NABU/prognos, Berlin. http://www.nabu.de/sanierungsfahrplan/NABU-Sanierungsfahrplan_endg.pdf. Accessed 21 Jan 2012

UBA (Umweltbundesamt) (2006) Wie private Haushalte die Umwelt nutzen—höherer Energieverbrauch trotz Effizienzsteigerungen: Hintergrundpapier November 2006. UBA, Dessau-Roßlau/Berlin. http://www.destatis.de/jetspeed/portal/cms/Sites/destatis/Internet/DE/Presse/pk/2006/UGR/UBA-Hintergrundpapier,property=file.pdf. Accessed 21 Nov 2011

Walberg D, Holz A, Gniechwitz T, Schulze T (2011) Wohnungsbau in Deutschland—2011: Modernisierung oder Bestandsersatz: Studie zum Zustand und der Zukunftsfähigkeit des deutschen „Kleinen Wohnungsbaus", Arbeitsgemeinschaft für zeitgemäßes Bauen e.V. Kiel. www.arge-sh.de. Accessed 3 March 2012

Weiss J, Dunkelberg E, Vogelpohl T (2012) Improving policy instruments to better tap into homeowner refurbishment potential: lessons learned from a case study in Germany. Energy Policy 44:406–415

Chapter 8
How Fuel Price Elasticity Affects the Economics of Thermal Retrofits

Abstract Policymakers, their expert advisors and the academic community use mathematical models to evaluate the economic viability and payback time of thermal retrofits. Most of these models have the form of a cost-benefit analysis, where the thermal upgrade costs are compared to the net present value (NPV) of the benefits expected to be received in future years, through fuel savings. However, these models assume that, if the dwelling had not been retrofitted, its occupants would have continued to consume the same amount of heating fuel as previously, despite future fuel price rises. In other words, they fail to include a factor for fuel price elasticity of demand. In this chapter, we show how the mathematics of these models can be modified to include this factor. We then test its effect by assessing the NPV and payback time of a set of thermal retrofit projects on a large housing estate in Germany. Even using conservative values for our parameters, the analysis shows that when price elasticity is taken into account NPV is reduced by around 23%, payback time is lengthened by 15–31 years, and the cost of abated CO_2 rises by around 27%.

Keywords Fuel price elasticity · Payback time · Economic viability · Cost-benefit analysis · Carbon abatement cost

8.1 Introduction

Throughout this book we have shown how occupant behaviour in the home is a major factor in determining actual, measured heating fuel consumption. In Chap. 5 we showed that German households consume 30% less heating fuel, on average, than the calculated energy ratings of their homes would indicate. In Chap. 7, we presented evidence that the reduction in domestic heating energy consumption in the years 2000–2009 appears to be largely caused by households becoming more thrifty with heating. This runs parallel to a steady increase in the price of heating fuel, averaging 4.8% per year, and there is evidence from studies in Germany

R. Galvin and M. Sunikka-Blank, *A Critical Appraisal of Germany's Thermal Retrofit Policy*, Green Energy and Technology, DOI: 10.1007/978-1-4471-5367-2_8, © Springer-Verlag London 2013

(Rehdanz 2007) and a number of comparable countries such as Denmark (Leth-Peterson and Togeby 2001) and Norway (Nesbakken 2002) that households respond to heating fuel price increases by consuming less.

If this is the case, it has implications for the economic viability of thermal retrofits. On average, households that do not thermally retrofit are likely to consume less heating fuel in future as the price of fuel rises. Yet this factor has not been taken into account in discussing the *economics* of thermal retrofits. If, for example, a household currently consumes 18,000 kWh per year on heating energy, then retrofits their property so that their consumption drops to 12,000 kWh per year, it is assumed this will save them 6,000 kWh per year, or 150,000 kWh over the 25-year lifetime of the retrofit measures. If the fuel price steadily rises over this time by, say, 4% per year, it is assumed that their payback through fuel savings will also rise by 4% per year. It is on this basis that calculations of economic viability are performed.

However, if this household had not retrofitted, we cannot in fact assume they would have continued to consume 18,000 kWh per year if the fuel price rose steadily at 4% per year. If, like the average household in Germany, they respond to heating fuel price rises by consuming less, they would have consumed less as the years went by. Hence, it is inaccurate to say they will save 150,000 kWh by retrofitting. In reality, they will save less.

German policymakers and their expert advisors use a well-known type of mathematical model to estimate economic viability and payback time of thermal retrofits: a 'cost-benefit analysis'. This compares the cost of the retrofit with the economic benefits it brings. If the benefits (accumulated over the technical lifetime of the retrofit measures) are equal to or outweigh the costs, the retrofit is economically viable. These models can also be inverted, to calculate the 'payback time' of a retrofit: how long it will take for the benefits to draw equal with the costs. The basic mathematical form of these models was introduced in Chap. 6, Sect. 6.4.

This chapter takes a closer look at these models and explores what happens if a factor is included in them to take account of this consumer sensitivity to fuel price rises.

A key concept throughout this chapter is 'heating fuel price elasticity of demand', which we often express more compactly, as 'price elasticity' or simply 'elasticity'. This is the percentage change in the quantity of fuel purchased (and hence consumed) for each 1% change in price. In Chap. 7, for example, we said that the average heating fuel price elasticity of demand for nonretrofitting households in 2000–2009 was −0.425. This means that for every 1% rise in fuel price in any particular year, these households, on average, purchased and consumed 0.425% less than they had previously.

In Germany, the factor of fuel price elasticity is absent from the cost-benefit models used to calculate economic viability and payback time. This can be seen, for example, in the models the Passive House Institute used in the reports they produced, commissioned by the Federal government, to confirm the economic viability of the thermal measures demanded in the EnEV in 2002 and 2009

(Feist 1998; Kah et al. 2008). It is also absent in models typically offered in countries such as Switzerland (Jakob 2006), the UK (Brechling and Smith 1992), Japan (Lopes et al. 2005), Greece (Papadopoulos et al. 2002), Denmark (Tommerup and Svendsen 2006), South Africa (Winkler et al. 2002) and Belgium (Verbeek and Hens 2005).

The challenge with developing a cost-benefit model for thermal retrofits is to include the appropriate parameters, with their appropriate values, in the mathematical formulas. Most current models include: the expected technical lifetime of the retrofit measures; the expected annual fuel savings; the expected annual future fuel price rise; and the discount rate (i.e. the annual fall in spending power of cash benefits that will be received in future years). These are all difficult to quantify and require some guesswork. However, homeowners have to make decisions as to whether, when and how deeply to retrofit, so the estimates offered by professionals are usually the best that these investors can hope to get.

Price elasticity is also difficult to estimate and it varies from household to household, but without it the models are skewed to the positive side: they overestimate the financial benefits of a retrofit.

We will show in this chapter how a factor for heating fuel price elasticity of demand can be incorporated in these models: how it modifies the equations and makes them, we argue, more realistic. In Sect. 8.2, we outline the basic mathematics of cost-benefit models for thermal retrofits and show how price elasticity modifies this. In Sect. 8.3 we calculate a value for price elasticity, based on empirical evidence presented in other chapters in this book. In Sect. 8.4 we apply this value, with the modified model, to an actual set of thermal retrofits in Germany. In Sect. 8.5, we invert our model to show how payback time is affected by price elasticity, and in Sect. 8.6 we consider what happens to the predicted cost of CO_2 savings when price elasticity is included in the calculations. In Sect. 8.7, we draw conclusions and discuss implications for policymakers and homeowners.

8.2 Including Price Elasticity in Cost-Benefit Analyses

8.2.1 The Basic Model

As we noted in Chap. 6, a thermal retrofit is considered economically viable if the total monetary savings expected to be gained from fuel savings are equal to or greater than the additional thermal costs. This can be expressed as:

$$C \leq B \text{ where}$$

$C = additional\ thermal\ costs$ (i.e. the costs of the thermal aspects of the upgrade, excluding any upgrade costs that do not improve the thermal quality of the dwelling);

B = monetary benefit, i.e. money saved through reduced fuel costs over the technical lifetime of the thermal renovation measures.

The monetary benefit B is worked out using a formula of the type:

$$B = Q_1 \times P_1 \times \frac{A^N - 1}{A - 1} \qquad (8.1)$$

where

Q_1 = quantity of fuel saved in the first year (kWh)
P_1 = price of each unit of fuel saved in year 1 (€/kWh)
N = number of years of the technical lifetime of the renovation measures
A = an 'annuity factor'

The expression: $\frac{A^N-1}{A-1}$ is the sum of the geometric sequence: $A^0 + A^1 + A^2 + \ldots A^{N-1}$, i.e. it adds up the savings made for all the years of the technical lifetime of the retrofit measures. The annuity factor A determines the way the present value of the benefits received in future years vary depending on the year they will be received. For $A > 1.0$ they steadily increase; for $A < 1.0$ they steadily decrease, and for $A = 1.0$ they remain constant. In the classic models A is made up of two factors: the expected annual fuel price rise and the discount rate:

$$A = \frac{1 + f/100}{1 + d/100}$$

where f = expected annual percentage fuel price increase and d = percentage discount rate. We can simplify this by setting:

$$F = 1 + f/100$$

$$D = 1 + d/100$$

Hence

$$A = F/D$$

Hence

$$B = Q_1 \times P_1 \times \frac{(F/D)^N - 1}{(F/D) - 1} \qquad (8.2)$$

The usefulness of a cost-benefit equation such as this is that all the benefits expected to be incurred and received in future years is translated into their equivalent values today, and added up to give total benefits: their 'net present value' (NPV). We should point out, however, that some economists use equations based on a different approach. They work out the *average* cost and benefit expected to be incurred and received in all the years of the technical lifetime of the renovations (e.g., Enseling and Hinz 2006).[1] Such models, however, are mathematically equivalent to the cost-benefit model introduced above and effectively produce the same results.

8.2.2 Including Price Elasticity of Demand

The year-on-year price elasticity of demand E is defined as the percentage change in the quantity Q_n of goods purchased in year n for each percentage change in fuel price P_n in year n. Hence:

$$E = \frac{(Q_{n+1} - Q_n)/Q_n}{(P_{n+1} - P_n)/P_n}$$

But $P_{n+1} = P_n \times F$ hence $(P_{n+1} - P_n)/P_n = F - 1$
Hence

$$E = \frac{(Q_{n+1} - Q_n)/Q_n}{F - 1}$$

Rearranging this equation we get:

$$Q_{n+1} = Q_n(E.F - E + 1)$$

$$\text{Let } H = E.F - E + 1 \tag{8.3}$$

Hence $Q_{n+1} = Q_n . H$, or more generally,

$$Q_n = Q_1 . H^{n-1} \tag{8.4}$$

H is a composite variable made up of the fuel price elasticity, which will have been derived from empirical studies of past years, and the expected fuel price rise during future years. It tells us the degree to which the quantity of fuel purchased falls each year, based on elasticity E and fuel price rise F. For example, if $H = 0.98$, this means that this year the consumer will purchase 98% of the fuel she purchased last year, and next year she will purchase 98% of this year's quantity, and so on—as long as the same annual fuel price rise continues. Hence, H enables us to project past purchasing responses to fuel price changes into the future. There are of course weaknesses in such projections, because there is no guarantee that people will act the same in the future as they did in the past. For example, their annual reductions in fuel consumption might reach saturation if their consumption falls below a certain level. On the other hand, their reductions in consumption

---

[1] In point of fact, Enseling and Hinz (2006) begin by stating the formula for such an analysis, but then proceed to perform a classic cost-benefit analysis of the type outlined here.

might accelerate if the fuel price reaches a certain threshold. So we proceed with these cautions in mind.

Using the formula for the sum of a geometric series, we can use Eq. (8.4) to derive a formula for the total quantity of fuel Q_{TN}, purchased in N years:

$$Q_{TN} = Q_1 \times \frac{H^N - 1}{H - 1} \tag{8.5}$$

We will return to Eq. (8.5) in our discussion of CO_2 savings, in Sect. 8.6. Meanwhile, we introduce the changing quantity of fuel purchased (Eq. 8.4) into our basic cost-benefit equation (Eq. 8.2).

Let K_n be the cost of fuel purchased in year n.

$$K_n = Q_n \cdot P_n$$
$$= Q_1 \cdot H^{n-1} \cdot P_1 \cdot F^{n-1}$$
$$= Q_1 \cdot P_1 \cdot (HF)^{n-1}$$

The present value of this cost will be K_n discounted by the discount rate:

$$K_{nd} = Q_1 \cdot P_1 \cdot (HF)^{n-1}/D^{n-1}$$
$$= Q_1 \cdot P_1 \cdot (HF/D)^{n-1}$$

Again using the formula for the sum of a geometric series, we find K_{TDN}, the total discounted cost of fuel consumed in N years:

$$K_{TDN} = Q_1 \cdot P_1 \cdot \frac{(HF/D)^N - 1}{HF/D - 1} \tag{8.6}$$

This applies to any home, retrofitted or not. To calculate the NPV of fuel that would be *saved* through retrofitting, we could calculate K_{TDN} for a nonretrofit scenario, then for a retrofit scenario, and the difference between these would be the NPV of fuel saved through retrofitting. More simply (but mathematically identically), we can set Q_1 as the fuel *saved* in year 1, to give B_{TDN}, the NPV of fuel saved over N years as a result of the retrofit, namely:

$$B_{TDN} = Q_1 \cdot P_1 \cdot \frac{(HF/D)^N - 1}{HF/D - 1} \tag{8.7}$$

This gives us the formula we need to calculate the NPV of a thermal retrofit, provided we know the year-on-year fuel price elasticity. We note that B_{TDN} is the 'benefit' in the cost-benefit analysis. This figure must be equal to or greater than the cost of the retrofit for it to be economically viable. We note that Q_1, the quantity of fuel saved in the first year, may be given in kWh or in kWh/m^2 of living area, as long as this is consistent with how the costs are given, i.e. in € or €/m^2 respectively.

8.2.3 Long-Run and Year-on-Year Fuel Price Elasticity of Demand

Empirical studies on fuel price elasticity of demand are usually conducted over several years and give one composite figure for the change in consumption over those years compared to the fuel price change in that time span (e.g. Rehdanz 2007). To use price elasticity in our formula, we need to translate this 'long-run' elasticity into 'year-on-year' elasticity. This is not as straightforward as it seems at first sight, as we are dealing with two exponential curves—one for fuel price and one for quantity purchased—that interact in different proportions depending on how far along the curves we are. Hence, we need to make the following mathematical transformation.

Let G = the long-run fuel price elasticity derived from an empirical study over n years. The definition of fuel price elasticity of demand over a particular period of time is the percentage change in the quantity of fuel purchased (and hence consumed), for each 1% change in price, over that period of time. Hence:

$$G = \frac{(Q_n - Q_1)/Q_1}{(P_n - P_1)/P_1} \tag{8.8}$$

We now substitute for Q_n from Eq. (8.2) and note that and $P_n = P_1 \cdot F^{n-1}$

$$G = \frac{(Q_1 \cdot H^{n-1} - Q_1)/Q_1}{(P_1 \cdot F^{n-1} - P_1)/P_1}$$
$$= \frac{H^{n-1} - 1}{F^{n-1} - 1}$$

We rearrange this equation to make H the subject:

$$G \cdot \left(F^{n-1} - 1\right) = H^{n-1} - 1$$

$$H = \sqrt[n-1]{1 + G \cdot (F^{n-1} - 1)}$$

But $H = E \cdot F - E + 1$ (from Eq. (8.3))
Hence $E \cdot (F - 1) + 1 = \sqrt[n-1]{1 + G \cdot (F^{n-1} - 1)}$

$$\text{So} \quad E = \frac{\sqrt[n-1]{1 + G \cdot (F^{n-1} - 1)} - 1}{F - 1} \tag{8.9}$$

This means that if we know the long-run elasticity G over n years and the annual percentage fuel price increase during those years f, where $F = 1 + f/100$, we can calculate the year-on-year fuel price elasticity (note that F refers here to the past years' fuel price increase, not that predicted for future years). If we then substitute E into Eq. (8.3) using the predicted fuel price increase for future years, this gives us H, the value we use in Eq. (8.7) to predict the NPV of fuel savings due to retrofits.

We now refer back to our analysis in Chap. 7 to establish a value for E, i.e. German year-to-year heating fuel price elasticity of demand.

8.3 Finding a Value for Fuel Price Elasticity

In Chap. 7, we estimated that heating fuel consumption among nonretrofitting German homes fell by 17.4% in 2000–2009. We also saw that heating fuel prices rose during this period by 52.5% in 2000–2009 (BMWi 2011a, b). This includes the different prices for natural gas, oil, coal, electricity, wood and other forms of biomass, weighted according to the quantity of each consumed nationally for domestic heating.

Recalling Eq. (8.8), for long-run fuel price elasticity G:

$$G = \frac{(Q_n - Q_1)/Q_1}{(P_n - P_1)/P_1}$$

Now

$$(Q_n - Q_1)/Q_1 = -17.4\%$$

And

$$(P_n - P_1)/P_1 = 52.5\%$$

Hence

$$G = 17.4/52.5 = -0.331$$

The average annual fuel price rise was 4.80%, since $\sqrt[9]{1.525} = 1.048$. Hence, $F = 1.048$ during this 9-year period. To translate G into year-on-year elasticity E, we use Eq. (8.9):

$$E = \frac{\sqrt[n-1]{1 + G \cdot (F^{n-1} - 1)} - 1}{F - 1}$$

$$= \frac{\sqrt[8]{1 + -0.331 \cdot (1.048^8 - 1)} - 1}{1.048 - 1}$$

$$= -0.425$$

This means that on average throughout 2000–2009, the average non-retrofitted household reduced its heating fuel consumption by 0.425% each year for every 1% annual increase in the price of heating fuel[2] We will now use this figure to see what difference this makes in a real, empirical situation.

---

[2] This assumes that all the reduction in consumption was a direct result of the fuel price increase. In fact, some of the reduction might be due to other factors, such as growing environmental awareness. A robust social science study would be needed to disaggregate these effects and produce a more accurate figure than the −0.425 used here. The present chapter is concerned principally with the mathematical modeling of price elasticity, using whatever figure is derived from empirical studies.

8.4 Fuel Price Elasticity in Practice

The chemical firm BASF (www.basf.de) developed polystyrene-based insulating materials in the early 1950s, and is Germany's largest manufacturer of these. Through its subsidiary company LUWOGE (www.luwoge.de), BASF owns large housing estates near its industrial base in Ludwigshafen-am-Rhein, which provide accommodation for employees. In 2002–2003, LUWOGE thermally retrofitted 150 of these apartment buildings, encompassing 850 apartments, with a total floor area of 48,000 m^2. The buildings had been constructed shortly after the World War II using thermally inferior materials, and in LUWOGE's estimation they were due for a comprehensive retrofit for reasons quite apart from their thermal quality.

After the retrofits, the Institut Wohnen und Umwelt (www.iwu.de) was commissioned by LUWOGE to assess and report on the economic efficiency of the thermal aspects of the upgrades. The report on this study (Enseling and Hinz 2006) is one of Germany's more frequently cited accounts of the costs and benefits of thermal retrofits. For example, the International Energy Agency reports the details of its findings as a normative example of what Germany is achieving in this sphere (IEA 2008). This makes it an interesting project on which to test our account of the influence of fuel price elasticity on the economics of thermal retrofitting.

The retrofits aimed to achieve four different thermal standards in different buildings: 150, 70, 40 and 30 kWh/m^2a. After the retrofits, consumption was measured, and found to have achieved averages, for each of the four groups, of 192, 70, 42 and 28 kWh/m^2a respectively.

Unfortunately, the records of actual, pre-retrofit consumption were not used in the assessment of how much heating fuel was saved through the retrofits. Instead, the pre-retrofit *calculated* consumption of 275 kWh/m^2a was used as a proxy for this. However, since economic viability and payback time have to do with actual fuel savings, we will not use this figure for our case study analysis. Due to the prebound effect (see Chap. 5), the actual consumption of homes with a calculated rating of 275 kWh/m^2a is likely to be around 180 kWh/m^2a. However, at the time these retrofits were undertaken national average consumption was some 10–15% higher than the period our prebound effect analysis covered, so we would estimate the average actual pre-retrofit consumption of these homes at around 210 kWh/m^2a.

Hence, we estimate the energy saved in each of these classes of retrofit as 18, 140, 168 and 182 kWh/m^2a respectively.

The report gives the 'additional thermal costs' (see Chap. 6, Sect. 6.2) as 36, 122, 187 and 314 euros per square metre of living area (€/m^2) respectively. Hence, the projects would qualify as 'economically viable' if the NPV of the fuel savings they brought per square metre of floor area, over the 25-year lifetime of the retrofit measures, was equal to or greater than each of these figures respectively. Recalling the equations derived above, to work out the NPV we need values for:

- Future fuel price rise: we will assume this continues at 4.8%
- Discount rate: we will assume this is 2.0%
- Cost of fuel in the first year after the retrofits: we set this at €0.05/kWh, in line with fuel prices at the time.

We will run the cost-benefit comparison twice for each of the four retrofit standards: the first time using the classic model that does not include fuel price elasticity and the second using our model which incorporates this. We note that in all these cases Q_1, the quantity of fuel saved in the first year, is given in kWh/m^2 and the costs of the retrofits are given in €/m^2.

8.4.1 Case 1. The Highest Standard (28 kWh/m²a) Without Fuel Price Elasticity

$$F = 1 + 4.08/100 = 1.048$$
$$D = 1 + 2.0/100 = 1.02$$

Hence $A = 1.048/1.02 = 1.0275$
Also $Q_1 = 182$ kWh/m^2; $P_1 = 0.05$ €/m^2
Using Eq. (8.1):

$$B = Q_1 \times P_1 \times \frac{A^N - 1}{A - 1}$$
$$B = 182 \times 0.05 \times \frac{1.0275^N - 1}{1.0275 - 1}$$
$$= 321 \ €/m^2$$

As we saw, the additional thermal costs C, of this class of retrofit were 314 €/m^2. So since $C < B$, this class of retrofit would be deemed economically viable.

We now see what happens when fuel price elasticity is included. Recalling that: $E = -0.425$ and $F = 1.048$, we use Eq. (8.3):

$$H = E \cdot F - E + 1$$
$$= -0.425 \times 1.048 - (-0.425) + 1$$
$$= 0.9796$$

This figure means that, if the homes had not been retrofitted and the annual fuel price rise continued at 4.8%, we could expect the average household to consume, each year, 97.96% of the quantity of fuel they consumed the previous year.

To find the effect of this we use Eq. (8.7):

$$B_{TDN} = Q_1 \cdot P_1 \cdot \frac{(HF/D)^N - 1}{HF/D - 1}$$

$$B_{TDN} = 182 \times 0.05 \times \frac{(0.9796 \times 1.048/1.02)^{25} - 1}{0.9796 \times 1.048/1.02 - 1}$$

$$= 246 \ €/m^2$$

The retrofit class would not be economically viable, as the benefit is 246 €/m² and the cost is 314 €/m². The inclusion of fuel price elasticity of demand has lowered the benefit by 23%.

8.4.2 The Other Three Cases

Table 8.1 gives the comparisons of benefits, with and without fuel price elasticity, for the other three cases and the above case. All these see a reduction in benefit of 23–25%. The case with the highest standard of retrofit qualifies as economically viable if fuel price elasticity is not included, but fails by 68 €/m² if price elasticity is taken into account. The case with the lowest standard almost qualifies as economically viable without fuel price elasticity, but fails by a large margin with fuel price elasticity accounted for. The other two cases qualify in both situations, but by a considerably smaller margin when fuel price elasticity is taken into account.

Here we have tested the effect of price elasticity in actual retrofit situations, using similar figures for fuel price, future fuel price rise and discount rate to those used by German policymakers. With these scenarios, fuel price elasticity makes a significant difference to the NPV of the fuel savings a homeowner can expect from a thermal retrofit project. Another way of considering the value of a retrofit is to calculate the payback time. In the next section we consider this approach.

Table 8.1 Comparison of accumulated savings over 25 years from four different classes of thermal retrofits, with and without the inclusion of fuel price elasticity of demand of −0.425

Retrofit standard (kWh/m²a)	Additional thermal costs (€/m²)	Energy saved in first year (kWh/m²)	Accumulated savings, without price elasticity (€/m²)	Accumulated savings, with price elasticity (€/m²)	Percentage fall in savings due to fuel price elasticity
28	314	182	321	246	23
42	187	168	296	227	23
70	122	140	247	189	23
192	36	18	32	24	25

Authors' calculations based on data from Enseling and Hinz (2006)

8.5 Price Elasticity and Payback Time

The payback time is the length of time it takes until the benefits from fuel saving draw equal to the cost of the retrofit, i.e. costs = benefits or $C = B$. We can invert Eq. (8.7) to find the year in which this occurs. Recalling Eq. (8.7) and substituting C for B:

$$C = Q_1 \cdot P_1 \cdot \frac{(HF/D)^N - 1}{HF/D - 1}$$

$$\Rightarrow (HF/D)^N - 1 = \frac{C(HF/D - 1)}{Q_1 \cdot P_1}$$

$$\Rightarrow HF/D = \sqrt[N]{1 + \frac{C(HF/D - 1)}{Q_1 \cdot P_1}}$$

$$\Rightarrow N = \frac{\log\left[1 + \frac{C(HF/D - 1)}{Q_1 \cdot P_1}\right]}{\log(HF/D)}$$

(8.10)

As a sample calculation we take the most energy-efficient retrofit, firstly without price elasticity (i.e. $H = 1$).

$$N = \frac{\log\left[1 + \frac{C(HF/D - 1)}{Q_1 \cdot P_1}\right]}{\log(HF/D)}$$

$$N = \frac{\log\left[1 + \frac{314 \times (1.048/1.02 - 1)}{182 \times 0.05}\right]}{\log(1.048/1.02)}$$

$$= 24.6 \text{ years}$$

We now re-run the calculation using fuel price elasticity, with $H = 0.9796$:

$$N = \frac{\log\left[1 + \frac{C(HF/D - 1)}{Q_1 \cdot P_1}\right]}{\log(HF/D)}$$

$$N = \frac{\log\left[1 + \frac{314 \times (0.9796 \times 1.048/1.02 - 1)}{182 \times 0.05}\right]}{\log(0.9796 \times 1.048/1.02)}$$

$$= 31.2 \text{ years}$$

Including fuel price elasticity of demand has lengthened the payback time from 24.6 to 31.2 years, or 27%.

The results for all four cases are displayed in Table 8.2. Inclusion of fuel price elasticity of demand increases payback time by 2.2 years, or 15%, for the third highest standard; 3.3 years (20%) for the second highest; 6.7 years (27%) for the highest standard; and 8.4 years (31%) for the lowest standard.

Table 8.2 Comparison of payback times of four different classes of thermal retrofit, with and without fuel price elasticity of demand of −0.425

Retrofit standard (kWh/m²a)	Additional thermal costs (€/m²)	Energy saved in first year (kWh/m²)	Payback time without fuel price elasticity (years)	Payback time with fuel price elasticity (years)	Percentage increase in payback time due to fuel price elasticity
28	314	182	24.6	31.3	27
42	187	168	17.6	20.9	20
70	122	140	14.4	16.6	15
192	36	18	27.4	35.8	31

Authors' calculations based on data from Enseling and Hinz (2006)

Because we have used a very conservative (low) value here for the discount rate, these increases are modest. A more realistic discount rate, of around 5%, would increase payback time considerably. For example, payback time would be infinite for the lowest standard and over 150 years for the highest standard. For more scenarios see Galvin and Sunikka-Blank (2012).

A further question is how price elasticity affects CO_2 savings and the cost of CO_2 abatement.

8.6 Price Elasticity and the Cost of CO_2 Savings

8.6.1 The Quantity of CO_2 Saved Through Retrofitting

The quantity of CO_2 saved as a result of a retrofit is directly proportional to the quantity of fuel saved (assuming there is no fuel switching). In the absence of fuel price elasticity of demand, the fuel saved for each of the 25 years of a retrofit's lifetime will be the same as the fuel saved in the first year. So, for example, for the highest quality retrofit the total CO_2 saved over the 25 years will be $25 \times 182\ \text{kWh/m}^2 = 4450\ \text{kWh/m}^2$. Using the conversion factor of 0.2 kg of CO_2 per kWh, this indicates that 0.910 tonnes of CO_2 will be saved for each m² of retrofitted living area (0.910 t/m²).

Expressing this more formally: recalling that Q_1 = quantity of fuel saved in the first year, we let Y be the quantity of CO_2 saved, in tonnes per m² of retrofitted living area, over N years:

$$Y = Q_1 \times N \times 0.2/1000 \qquad (8.11)$$

To find how fuel price elasticity will affect this, we note that the quantity of fuel saved is reducing every year by the factor H, derived from the fuel price elasticity. Hence:

$$Y = Q_1 \times 0.2/1000 \times \left(H^0 + H^1 + H_2 + \ldots H^{N-1}\right)$$

Noting that the expression in brackets is the sum of a geometric series, the formula becomes:

$$\Rightarrow Y = Q_1 \times 0.2/1000 \times \frac{H^N - 1}{H - 1} \tag{8.12}$$

For the highest quality retrofit this will be:

$$Y = 182 \times 0.2/1000 \times \frac{0.9796^{25} - 1}{0.9796 - 1}$$
$$= 0.718\,\text{t/m}^2$$

This indicates that fuel price elasticity reduces the quantity of CO_2 saved as a result of the retrofit, in this case by 21%.

For the other three cases, the reductions in fuel saved will be in the same proportion, since H is the same in all cases.

But this does not imply that less CO_2 will be saved if we retrofit than if we do not retrofit. Rather, it implies that the quantity of CO_2 saved as *a result of retrofitting* will be 21% less than expected. The 21% would have been saved if we had not retrofitted, so crediting it to the retrofit is an accounting error. The omission of price elasticity from the basic cost-benefit model misleads us into thinking the retrofit is the cause of more CO_2 savings than it actually is.

This leads to the issue of the cost of CO_2 saved.

8.6.2 The Cost of CO_2 Saved Through Retrofitting

For the highest thermal standard, considered above, without fuel price elasticity, we found that 0.910 tonnes of CO_2 are saved for each m^2 of retrofitted living area. The additional thermal costs—i.e. the cost of saving this CO_2—were 314 €/m². Hence, the cost of CO_2 abatement was 314/0.910 = €345/tonne. For the same thermal standard, but with fuel price elasticity taken into account, the cost is 314/0.717 = €438/tonne, an increase of 27%. Table 8.3 lists the costs of CO_2 abatement for all four retrofit standards, with and without price elasticity.

Here we see that, without price elasticity, the cost of CO_2 saved ranges from 174 €/t for the third best thermal standard to 400 €/t for the least best standard, with that for the best standard 345 €/t. When price elasticity is included, these costs range from 221 to 507 €/t, with the highest standard 438 €/t. Including fuel price elasticity of demand increases the cost of CO_2 abatement by 24–27% in these cases.

Including fuel price elasticity in our calculation reveals that by retrofitting to such high thermal standards, we are shifting the source of a significant part of the payment for CO_2 savings from the free savings won through consumer response to fuel price increases to the very expensive savings achieved through top-end

Table 8.3 Quantity of CO_2 saved over 25 years, and cost of CO_2 saved, for each retrofit standard, without and with fuel price elasticity of demand

Retrofit standard (kWh/ m^2a)	Additional thermal costs (€/m^2)	Energy saved in first year (kWh/m^2)	Without price elasticity		With price elasticity		% increase in cost of CO_2 saved
			Quantity of CO_2 saved (t/m^2)	Cost of CO_2 saved (€/t)	Quantity of CO_2 saved (t/m^2)	Cost of CO_2 saved (€/t)	
28	314	182	0.910	345	0.718	438	27
42	187	168	0.840	227	0.662	282	24
70	122	140	0.700	174	0.551	221	27
192	36	18	0.090	400	0.071	507	27

Authors' calculations based on data from Enseling and Hinz (2006)

thermal retrofitting. This is not to deny the value of top-end thermal retrofits, but merely to point out that the accounting has to be done properly: a good portion of the CO_2 savings that occur after top-end retrofits would have occurred anyway, and therefore cannot be credited to the retrofit measures.

8.7 Conclusions and Implications

This chapter set out to fill a gap in the cost-benefit models used to calculate the NPV and payback time of thermal retrofits of homes in Germany. This gap is the factoring in of price elasticity of demand for heating fuel. We developed equations for incorporating this factor into the basic form of these models. Mathematically, the inclusion of fuel price elasticity of demand is bound to reduce the NPV of the fuel saved through the retrofit and lengthen its payback time, as it reduces the annuity factor that determines the NPV. By applying both the traditional model and our new model to a set of real thermal retrofits in a large retrofit project, we found that the reduction of NPV and the lengthening of payback time through price elasticity are of significant magnitude. In the scenarios we offered, NPV is reduced by around 23%; payback time is increased by between 15 and 31%, and the cost of CO_2 saved is increased by 24–27%.

These results are conservative, as we chose a conservative (i.e. low) value for the discount rate in the cost-benefit formulas. A higher discount rate would reduce the present value of future fuel savings by a much greater factor, and in some cases this would not only make a retrofit no longer economically viable; it could also lead to payback times being 'infinite'. This is because the present value of each year's fuel savings would diminish rapidly as the years progress. A high (or some would say realistic) discount rate of 8% or more, which is common among commercial UK housing providers, would render the present value of payments expected to be received in 15 or more years very low. The mathematical explanation for this is that with a high value of D or a low value of H, the factor $H \cdot F/D$ would be sufficiently low to produce a steeply convergent geometric sequence for the present values of future savings in fuel, and the sum to infinity of a convergent geometric sequence is a finite number.

A further important issue is that, when fuel price elasticity is included, the assumption that future fuel price rises will be high does not necessarily make retrofit projects more economically attractive. For high annual percentage fuel price rises, the factor H becomes low, since fuel price elasticity reacts with higher fuel price rises to produce greater reductions in fuel consumption, and therefore lower annual fuel savings through retrofitting. If H is low, the factor $H \cdot F/D$ is low, and therefore the net present benefit is low.

Even under the most ideal conditions for a retrofit scenario, fuel price elasticity reduces the NPV of the fuel savings and lengthens payback time. For policy-makers, this puts a further question mark over the promoting of thermal retrofits on the grounds that they are 'economically viable' for homeowners.

This is not to deny the value of top-end thermal retrofits for other reasons than economic gain. If homeowners want to escape the constraints of behaving more fuel-thrifty as fuel prices rise, or if they want to retrofit for reasons other than to save money, they may be encouraged to go ahead and pay the price. People with high incomes renovate kitchens and bathrooms for reasons other than financial gain, to keep their home comfortable and technologically advanced. The government, we suggest, should promote top-end thermal retrofits based on these reasons, not on the grounds that they are a good way to save money.

References

BMWi (Bundesministerium für Wirtschaft und Technologie) (2011a) Energiedaten: ausgewählte Grafiken; Stand: 15.08.2011a. http://www.bmwi.de/BMWi/Navigation/Energie/Statistik-und-Prognosen/energiedaten.html Accessed 14 Dec 2011

BMWi (Bundesministerium für Wirtschaft und Technologie) (2011b) Zahlen und Fakten: Energidaten: Nationale und Internationale Entwicklung, Bundesministerium für Wirtschaft und Technologie, Referat III C 3. BMWi, Berlin. http://www.bmwi.de/BMWi/Navigation/Energie/Statistik-und-Prognosen/energiedaten.html. Accessed 5 Dec 2010

Brechling V, Smith S (1992) The pattern of domestic energy measures among domestic households in the UK. Institute for Fiscal Studies, London

Enseling A, Hinz E (2006) Energetische Gebäudesanierung und Wirtschaftlichkeit – Eine Untersuchung am Beispiel des 'Brunckviertels' in Ludwigshafen. Institut Wohnen und Umwelt GmbH, Darmstadt

Feist W (1998) Wirtschaftlichkeitsuntersuchung ausgewählter Energiesparmaßnahmen im Gebäudebestand, Fachinformation PHI-1998/3. Passivhaus Institut, Darmstadt

Galvin R, Sunikka-Blank M (2012) Including fuel price elasticity of demand in net present value and payback time calculations of thermal retrofits: case study of German dwellings. Energy Build 50:219–228

IEA (International Energy Agency) (2008) Promoting energy efficiency investments: case studies in the residential sector. Int Energy Agency, Paris

Jakob M (2006) Marginal costs and co-benefits of energy efficiency investments: the case of the Swiss residential sector. Energy Policy 34:172–187

Kah O, Feist W, Pfluger R, Schnieders J, Kaufmann B, Schulz T, Bastian Z, Vilz A (2008) Bewertung energetischer Anforderungen im Lichte steigender Energiepreise für die EnEV und die KfW-Förderung, BBR-Online-Publikation 18/08. BMVBS/BBR, Herausgeber. http://www.bbr.bund.de/cln_015/nn_112742/BBSR/DE/Veroeffentlichungen/BBSROnline/2008/ON182008.html. Accessed 10 Nov 2011

Leth-Peterson S, Togeby M (2001) Demand for space heating in apartment blocks: measuring effects of policy measures aiming at reducing energy consumption. Energy Econ 23(4):387–403

Lopes L, Hokoi S, Miura H, Shuhei H (2005) Energy efficiency and energy savings in Japanese residential buildings—research methodology and surveyed results. Energy Build 37:698–706

Nesbakken R (2002) Energy consumption for space heating: a discrete–continuous approach. Scand J Econ 103(1):165–184

Papadopoulos A, Theodosiou T, Karatzas K (2002) Feasibility of energy saving renovation measures in urban buildings the impact of energy prices and the acceptable pay back time criterion. Energy Build 34:455–466

Rehdanz K (2007) Determinants of residential space heating expenditures in Germany. Energy Econ 29:167–182

Tommerup H, Svendsen S (2006) Energy savings in Danish residential building stock. Energ Buildings 38:618–626

Verbeek G, Hens H (2005) Energy savings in retrofitted dwellings: economically viable? Energ Buildings 37:747–754

Winkler H, Spalding-Fecher R, Tyani L, Matibe K (2002) Cost benefit analysis of energy efficiency in urban low-cost housing. Dev S Afr 19(5):593–614

Chapter 9
Conclusions: A New Way Forward

Abstract Germany has engaged better-off homeowners in advanced thermal retrofits to high standards. However, at this level retrofits are inherently economically inefficient, and empirical research shows they often bring significantly smaller savings than calculated. This 'top-end' approach also fails to engage the bulk of homeowners, due to severe technical and economic difficulties of retrofitting to the required level. In order to increase the annual retrofit rate and energy savings, we propose a broadening out of policy into three distinct streams, simultaneously promoting: *cost-effective* thermal upgrade measures, which would not necessarily meet current, stringent EnEV standards; *user behavior* change, which appears to have a very large saving potential that is already beginning to be realized; and *top-end* thermal retrofits which may not be economically viable but could be promoted for their environmental and comfort benefits. To put this three-stream CUT model into place would require institutional change, and also a significant shift in attitude among policymakers. They would have to accept the value of modest thermal upgrade measures, and of the contribution of behavior change, rather than remain narrowly focused on the extreme end of technological capability. Notions of 'economic viability' would also need to be overhauled.

Keywords Retrofits · Thermal regulations · German policy · Economic viability · Energy use behavior

9.1 Introduction

This book has critically analyzed the German project of thermal retrofitting of existing homes. We have found much of value in this policy, but also seen that the results it is achieving fall far short of what policymakers have aimed for. The policy is leading to real, measurable reductions in domestic heating fuel consumption of about 0.25% per year, whereas this needs to be about 2.1% to reach the policy goal of 80% reductions by 2050.

R. Galvin and M. Sunikka-Blank, *A Critical Appraisal of Germany's*   135
Thermal Retrofit Policy, Green Energy and Technology,
DOI: 10.1007/978-1-4471-5367-2_9, © Springer-Verlag London 2013

In this chapter we summarize the reasons for this shortfall, but also take the matter one step further. We believe our detailed analysis has not only uncovered weaknesses of this policy, but has also brought to light pointers as to how it can be improved. We think a set of well-designed policy initiatives along certain lines could lift the current 0.25% annual consumption reduction rate to well over 1%, without any significant increase in government spending. Much of this chapter is therefore devoted to outlining our suggested approach to accelerate heating fuel savings from the existing German homes. We call this the CUT approach, as it is based on a balance of: Cost-effective retrofit measures; User behavior strategies; and Top-end retrofit measures. We will be suggesting that the German Government has depended far too heavily on the third of these. It has pursued unbalanced policies, investing almost all its energies and legal powers in promoting expensive, technically difficult 'deep' retrofits that are often individually impressive but bring relatively little return per euro invested. In so doing, it has narrowed the appeal and accessibility of thermal retrofits to a small segment of well-off private homeowners and well-funded housing providers. Meanwhile, it has neglected other potential fuel and CO_2 savings that are waiting to be tapped, and that are found across the board in the German housing stock.

We will propose an alternative policy approach, which we hope will find fertile ground among Germany's policymakers—many of whom we have shared discussions with in the course of our research. First, however, we will summarize the results of our research as outlined in this book. This summary is given in Sect. 9.2. In Sect. 9.3, the CUT model is outlined in practice. Final concluding remarks are offered in Sect. 9.4.

9.2 Summary of Findings

9.2.1 The Current Policy

Germany has taken bold steps to improve the energy efficiency of its housing stock over the last decade. The Energy Saving Regulations (EnEV), brought into force in 2002 and strengthened in 2009, require homeowners to thermally upgrade their properties to specific standards when they do regular or incidental maintenance, repairs, or extensions. The maximum permissible theoretical (calculated) heating energy consumption for a comprehensively renovated building is an average of 100 kWh/m²a, while that for individual measures, such as a wall or roof refurbishment, are the equivalent of a building designed to reach 70 kWh/m²a. This compares with an average theoretical standard in the current building stock of around 225 kWh/m²a.

The average depth to which homes have been thermally retrofitted increased steadily from the mid 1990s to the late 2000s, but has now reached a plateau. The limits of technically feasible, economically efficient thermal renovation now appear to have been reached, and this is reflected in the government's recent decision not to tighten the regulations further.

Germany's thermal retrofit policy is closely tied to its climate goal of 80% reductions in CO_2 emissions compared to 1990 levels by 2050. Despite progress made since 1990 in new build and thermal retrofit technology, in 2013 Germany is still left with a housing stock that would need its entire heating fuel consumption to be reduced by 80% for the climate goal to be reached in respect of this sector.

The average *theoretical* reduction being achieved in Germany's retrofit projects—including those receiving subsidies from the German Development Bank (KfW) for attempting to go beyond the EnEV requirements—is a disappointing 33%. The *actual* savings, based on measured pre- and post-retrofit consumption is even lower, at around 25%. The equivalent of around 1% of total living area is being retrofitted to this standard annually, putting the national annual reduction in energy consumption, through thermal retrofits, at about 0.25%. If this rate continues for the 38 years from 2013 to 2050, it will produce a saving of 9.5%. This is a long way from the required 80%.

The reasons this progress is slow can be discussed under three broad headings, though they are deeply intertwined: technical, economic and user-behavior challenges.

9.2.2 Technical Challenges

The biggest technical challenge is the 'law of diminishing returns'. A modest depth of insulation can provide relatively large fuel savings for a thermally poor dwelling, but as the thermal quality improves, it takes proportionately greater depths of insulation to achieve significantly larger savings. This is a physical, mathematical reality, which applies even in ideal shaped buildings. But actual existing homes present additional, practical difficulties: roof overhangs that are too short to accommodate 16 cm of wall insulation; basements that become too shallow with 10 cm of insulation attached to the ceiling; wall insulation that protrudes into driveways, balconies, and neighbors' properties; windows that seem to disappear down dark tunnels when wall insulation is too thick; roofs that need to be rebuilt to accommodate 22 cm of insulation under the tiles. Older homes also often have many corners, attached balconies and gables. These increase the outside surface area, thereby increasing the heat loss, and also present difficulties for attaching thick insulation. They can also bring conflicts between architectural values and measures to improve energy efficiency.

Further, the tight air seal required for high thermal standards leads to the need for good ventilation, but building a mechanical ventilation system into the structure of an old home is technically difficult and expensive. Without such a system occupants have to ventilate manually—which can lead to large wastage of heating energy—to ensure good indoor air and avoid mould growth. Challenges such as these have led to homeowners, academics and the construction industry giving increasing feedback to the government that the technical requirements of the EnEV are already too strict for many projects.

9.2.3 Economic Constraints

The second set of difficulties is economic. In Chap. 6, we saw that the EnEV sets thermal retrofit standards at the limit of what policymakers believe to be 'economically viable'—where the fuel savings over the technical lifetime of the retrofit measures are expected to pay for the costs of those measures. However, we identified six key problems with this approach. First, it only applies when a building is being retrofitted anyway, for regular cyclical maintenance, since only the 'additional thermal costs' are counted in the calculation of economic viability. Second, even for such cases as these, the technical difficulties outlined above can lead to step-wise cost increases. Third, the economic viability criterion assumes a pre-retrofit heating fuel consumption level equal to the calculated energy rating, whereas actual consumption is, on average, 30% below this—what we call the prebound effect. This drastically reduces the actual fuel savings achieved through thermal retrofits, as occupants cannot save fuel they were not already consuming.

Fourth, even if this were not a problem, the economic viability criterion depends on an exponential curve of future fuel price rises, so that investors suffer cumulative losses for some 14 years before the returns begin to pare back these losses. Many homeowners do not like investing over such long time horizons, and this reduces the appeal of thermal retrofits as a financial investment. Fifth, the economic viability models fail to take into account price elasticity of demand for heating fuel (Chap. 8). This reduces the return on a retrofit investment by around 25% and can lengthen payback time from a few years to many decades. Finally, the marginal costs of thermal retrofits to EnEV standards represent CO_2 abatement costs of €250–€1000 per tonne of avoided CO_2 emissions. This compares with typical costs in other sectors of less than €30, making retrofitting to EnEV standards a very economically inefficient way to achieve CO_2 abatement goals even in cases where it is economically viable.

These economic difficulties, we believe, are the main reason the annual rate of thermal retrofits has stalled at around 1%.

9.2.4 Challenges Presented by User Behavior

The way occupants behave in relation to their day-to-day home heating is one of the reasons for the low magnitude of fuel savings being achieved through thermal retrofits. To begin with, the prebound effect means that often there is not as much savings potential in a retrofit as expected. Further, homeowners who were fuel thrifty prior to a retrofit might feel they deserve to be warmer after spending large sums on the retrofit, and respond by keeping higher indoor temperatures. If these go above the 19 °C assumed in the calculated energy rating, the prebound effect may go into reverse, so that households consume more energy after a retrofit than the post-retrofit rating.

On the other hand, the prebound effect has the positive advantage of bringing heating energy savings *without* retrofitting. We saw (Chap. 7) that these savings appear to have been increasing steadily over the last decade. This could be partly a response to rising fuel prices (Chap. 8), or to demographic and lifestyle trends, or to increased ecological concern. We will return to this issue in our discussion of our CUT proposal below.

A further behavior-based reason for the low savings rate through thermal retrofits is that German occupants are not well skilled at ventilating their homes energy efficiently, a problem that becomes more acute for thermally retrofitted dwellings. A case study carried out in Aachen, a medium-sized city in northwest Germany, revealed that energy-efficient ventilation is the exception rather than the norm (Galvin 2013). To solve the problem of moisture build-up in air-tight dwellings, most households ventilate by putting windows on the trickle ventilation setting (tilted open at about 10° from the vertical) for several hours a day. This sends warm air out the window, cools the interior substance of the building, and can consume up to 30 kWh per day. The energy-efficient way to ventilate such homes is by 'shock-ventilation' *(Stoßlüften)*, in which all windows and internal doors are opened completely, for 2–3 min only, several times per day, and kept firmly shut the rest of the time. Each shock-ventilation consumes around 1 kWh, a fraction of consumption by trickle ventilation. The prevalence of trickle ventilation in Germany is no doubt a major reason for the low savings rate through thermal retrofits, and deserves to be more thoroughly investigated.

These types of factors—technical, economic, and behavioral—appear to be combining to make for a disappointingly low rate of annual reductions in energy consumption through thermal retrofits. Our most recent discussions with policymakers in Berlin indicate that there are two broad streams of thought as to what to do about this. Some are saying we must simply push harder: tighten the thermal standards further; increase the level of subsidies for top-end retrofits; legislate to require homeowners to thermally upgrade their properties even if not doing maintenance; introduce a strict inspection scheme with powers to impose heavy fines for noncompliance; pull down more old buildings and replace them with 'nearly zero energy' homes (for examples of this approach see BMWi/BMU 2010, 22ff; Tschimpke et al. 2011). We see this as impractical and economically inefficient. It would divert large quantities of money and expertise to top-end upgrades which give a low return on energy saved per euro invested and incur huge CO_2 abatement costs, not to mention possible human rights issues of forcing homeowners to pay for most of it.

Other policymakers, however, are hoping there might be some other way forward. We now introduce an approach that should cost no more Federal money than the present one but could unlock yet-unrealized domestic heating fuel saving potential, lifting the annual rate of reduction in heating energy consumption by a very significant proportion.

9.3 A Way Ahead: The CUT Model

The policy approach we are suggesting for thermal retrofits is based on a pragmatic balance of: Cost-effective retrofit measures; User behavior strategies; and Top-end retrofit measures. We propose that Federal policy needs to be re-thought and re-developed, so that all three of these are emphasized and promoted equally. This would require certain changes to the EnEV, but equally important, it would need three parallel, focused promotional and enabling thrusts, each well funded, and critically informed by ongoing, interdisciplinary research.

9.3.1 'C': Cost-Effective Thermal Upgrade Measures

We have argued that retrofits to EnEV standards are not, for a majority of homes, the benignly economically viable projects they are made out to be. However, we have also argued that some thermal retrofit measures generally are economically viable (Chap. 6), and that if these are done sensibly they can also be economically *optimal*—they can bring the greatest possible return in energy saved per euro invested. For example, a layer of 12 cm of insulation attached between the rafters of a roof (i.e. in a common accessible attic in an apartment building) might not only bring a positive monetary return from fuel savings, but also save more fuel per euro invested than any other thickness, larger or smaller. For a different roof, the economic optimum might be 10 or 16 cm or, if the loft is not used as accessible space, the economical optimum might be to lay 25 cm of insulation on the loft floor. There might also be an economically optimal thickness of insulation for the basement ceiling if the basement is shallow. Other examples might be filling cavity walls, installing hydraulic equalization in the central heating system, or applying air-sealing strips to stop draughts through the windows.

In some cases, wall insulation can also be both economically viable and economically efficient. If a wall needs major repairs or maintenance, a certain thickness of external wall insulation might pay back through fuel savings within a decade. But this may have to be less than the EnEV standard of around 16 cm to avoid roof realignment or repositioning of driveways etc. Alternatively, internal wall insulation can be far cheaper than external, but to preserve indoor living space this usually has to be much thinner than 16 cm. For example, a 4 cm layer, correctly applied with moisture avoidance gaps, can make a significant saving. It can reduce the U-value of a section of wall from a chilly 3.5–1.0 W/m^{2}K, a reduction in heat loss of 71%. This is not as thermally effective as an EnEV-standard 16 cm layer, which would bring a U-value of 0.25 W/m^{2}K and a reduction for that section of wall of 94%, but is far cheaper and many times more economically efficient.

A thermal upgrade measure which could work on a very large scale in Germany is roof insulation. The EnEV requires roofs to be insulated to achieve a U-value of 0.20 W/m²K or less. This requires 20–24 cm of insulation. Since the rafters are generally only 12 cm deep, there are two main options to achieve this. The cheaper of these is to insulate both between and under the rafters, though this requires highly skilled labor and reduces the useable area of the loft considerably. The more expensive option is to insulate from outside: either rebuild the roof, or remove the tiles, fix a layer of hard insulation material on top of the rafters and place new tiles on top of that. This would require professional contractors and cost around €20,000 for a typical apartment building (Simons 2012, p. 19).

The cost-effective option, however, is often to fit a 12 cm layer of insulation between the rafters internally. Interestingly, the EnEV now has additional clauses, in appendices, that allow for this option if the existing building structure makes the standard 22 cm depth technically unworkable (GdW 2010). This is an example of how regulations for the existing stock can in fact be attuned in a more nuanced way to accommodate the characteristics of buildings and the economic restraints homeowners are under. However, because this is seen as a technically inferior option, it is not promoted by the government and our enquiries revealed that local authorities are often not even aware it exists as a legal option.

Nobody knows how many of Germany's 18 million roofs of residential buildings are uninsulated, but from our discussions with German Energy Agency we estimate this to be around 10 million. If 260,000 of these were insulated internally in a cheap, modest, economically optimum way each year, this could easily save 2,000 kWh per year for each building, reducing Germany's heating consumption by a further 0.52 TWh per year, or 0.11%. In 38 years, this would bring national reductions of 4.5%.

Based on a particular, typical apartment building roof in Aachen and current prices, we estimate that a problem-free job of this type would cost about €600. At the current fuel price of €0.10/kWh, this would pay back, through fuel savings, in 3 years. If at any time after that the tiles are ready to be replaced and it becomes economical to fit external roof insulation, there would be no net monetary loss from insulating twice (i.e. it would not cause 'lock-in' to a low standard). This genuinely easy gain is currently available, but the government does not promote it.

Another pair of cost-effective, easy gains are boiler adjustment and hydraulic equalization for distributing heat more evenly in multi-storey buildings. These cost little and can bring significant savings, and are already promoted, to some extent, by government campaigns and the literature. We are suggesting that a special branch of policy needs to focus on promoting cost-effective thermal upgrade measures, regardless of how technically interesting (such as hydraulic equalization) or commonplace (such as insulation placed between rafters) they might be.

What would have to change, for such a policy to be embraced? First, we suggest that a prevailing discourse of technological perfectionism would need to loosen its hold. Interviews with policymakers and their expert advisors at Federal, state, and municipal level have revealed a firm attachment to the notion that only the best technology is good enough: if we now have the know-how to attach 16 cm of external wall insulation, we must do it. Phrases such as *'es muss richtig sein'* (it must be done correctly) and *'wenn schon, denn schon'* (if a job's worth doing, it is worth doing properly) were frequently uttered in defence of the tight standards demanded in the EnEV (Galvin 2011; see also Simons 2012, p. 20).

In contrast, a researcher at GdW, the Association of German Housing Providers (www.gdw.de) pointed out that 15 years ago policy actors were excited that it had then become possible to attach layers of 8 cm of external wall insulation. 8 cm, she said, was 'all the rage' *(der letzte Schrei)*. As thicker layers began to be used, however, 8 cm became passé, then substandard, then in most circumstances illegal. Even though there are severe economic and technical problems in attaching 16 cm, which make many homeowners shy away from retrofits, this was lauded as the correct standard that must be adhered to.

We suggest Fritz Schumacher's phrase 'appropriate technology' (Schumacher 1973) would be more suitable for the 'C'-strand of policy than technological perfectionism, which can be confined to top-end renovation measures of the 'T'-strand. The Housing Ministry (BMVBS), responsible for the EnEV, has developed a tight regime of promoting top-end thermal retrofits and discouraging attempts to retrofit at more modest levels. The quasi-independent German Energy Agency (DENA) has acted similarly, reaching into the community with exemplar retrofits, promotional literature and frequent press statements. The Ministry of the Economy and Technology (BMWi) and the Ministry of the Environment (BMU) have also made clear their view that top-end retrofits are the only acceptable standard (BMWi/BMU 2010). These bodies would need to shift gear so as to accept and accommodate the 'C' branch of a new policy initiative—not instead of but alongside retrofits at the top-end standard.

A second change of discourse would be an emphasis on the 'economic optimal' rather than on the outer limit of what is 'economically viable'. There is now a large body of research on the economic optimal in energy and CO_2 saving, and the EU Commission is now taking this issue seriously (Boermans et al. 2011). A case could be made for re-wording the EnEV to require those who retrofit to reach at least an economically optimum level of energy savings. Hence, instead of using 'economic viability' as a central motivating tool, the 'C' branch of policy would emphasise economically optimum retrofit measures, backed up with a flexible approach from energy advisors. Germany currently has no inspection regime for building upgrades, but we suggest that a combined advisory and inspection service could be formed out of its current home energy advisor networks to support these retrofit measures.

9.3.1.1 Saving Potential from 'C' Stream

Under the strict standards of the EnEV, around 1% of dwellings are thermally upgraded each year, producing average actual gains of around 25% per upgraded dwelling. If there were an energetic policy focus on economically optimum measures in this 'C' stream, we suggest this could produce an annual retrofit rate, at this modest standard, of up to 3%, while possibly producing actual savings of around 15% per upgraded dwelling. This would provide annual national savings of 0.45%. Continuing at this rate would exploit all such opportunities in the housing stock in 33 years, i.e. by 2045, and bring total savings of around 15%.

9.3.2 'U': User Behavior

In this book, we have reviewed the evidence that user behavior contributes at least as much to the quantity of heating energy consumed, as does the technical quality of buildings. Research by the Environment Ministry (UBA) a decade ago found strong empirical evidence for this in Germany (UBA 2006). The EU Commission has recognized that household behavior change '...can result in large reductions of greenhouse gas (GHG) emissions in the EU, both in the shorter and in the long term' (Faber and Schroten 2012). Qualitative studies in European countries are now revealing some of the key dynamics in this (e.g. Gram-Hanssen 2010). Our own work on the prebound effect (Chap. 5), on the fall in consumption in 2000–2009 (Chap. 7) and on ventilation practices (Galvin 2013) concurs with these findings and enables us to begin to quantify the effects of user behavior on heating consumption.

Many German households are already saving significant quantities of energy by adopting more economical day-by-day heating behavior. Average consumption is 30% below the calculated ratings of dwellings, and this figure masks a wide spread of differences between households. Some are consuming much less than this, others more. We have also found that, on average, this downward trend is increasing. We calculated that in the years 2000–2009, heating consumption in older, non-retrofitted homes not only fell by around 17.4%, but also made up over 80% of the total fall in Germany's heating fuel consumption in those years.

These findings might point to some degree of fuel poverty in Germany, but their extent and magnitude suggest that energy-saving user behavior is commonly practiced in Germany and is becoming more widespread, without large-scale complaints of cold, uncomfortable homes.

We have good knowledge of the strategies occupants can use to reduce their heating energy consumption while remaining comfortable in their homes. The three most obvious ones are: heat only the rooms being used, wear sensibly warm clothes at home so as to keep lower indoor temperatures; and ventilate non-wastefully. In a test apartment in Aachen, a household following our advice applied these strategies in a disciplined way for a full year, without any reported

loss of comfort. The apartment's calculated energy rating was 124 kWh/m²a
(based on 'useable' area), but the actual consumption over the year was 36 kWh/
m²a—a saving of 74%. This is significantly lower annual consumption than even
the new-build standard of 70 kWh/m²a.

But although we know the strategies that work, we do not know why some
households implement them while others do not. There is increasing evidence that
household energy-saving behavior is strongly determined by people's desire and
willingness to save energy, and also by several other factors (Hargreaves et al.
2010). The main ones are: households' daily routines and how these fit with or
compromise attempts to save energy; the discourse, or subtle messages associated
with energy saving among household members; the skills occupants possess to
manipulate technical systems; and the user-friendliness of the heating system's
adjustment interface (Shove 2010). Attempts to induce occupants to develop
energy-saving behavior need to address all these issues together. It is not simply a
matter of giving out leaflets on smart ways to save heating energy, or of installing
smart meters in homes. We suggest that a branch of policy needs to be developed
that researches and implements effective ways to address these issues, and so to
harness the large, untapped potential savings through user behavior change.

The Federal Environment Agency (UBA) has a record of proposing ambitious
CO_2 reduction goals (e.g. Flasbart 2009) and has long promoted user behavior
change as a means of reducing domestic heating energy consumption. This has not
come across to the public with much force, but it represents the only long running,
significant Federal initiative to induce user behavior change for household heating.
Perhaps this Agency could be the one to head up a concerted, informed, well-
resourced approach to engage households in fuel saving strategies. An early initiative could be a social science-based search for exemplar households that are
successfully and comfortably living with low heating energy consumption in non-
retrofitted homes. This would help policymakers better understand what motivations, attitudes, discourses, routines, skills and physical environment are associated
with such savings. Findings such as this could lead to behavior models that might
be able to be reproduced among other households of various types.

9.3.2.1 Saving Potential from 'U' Stream

The 17.4% reduction in heating fuel consumption through behavior change in
German households in 2000-2009 represents an annual reduction of 1.8% (see
Chap. 7). If these reductions could be induced to continue at only a quarter of this
rate for the next 38 years they would bring a saving of 17%. If all possible savings
of this kind were exhausted after 28 years, the saving would still be a healthy 13%.
We believe that with robust and consistent policy support such a goal could well
be achievable.

9.3.3 'T': Top-End Thermal Retrofitting

The thermal upgrade measures demanded by the EnEV can be very effective in saving heating energy, even if they do not pay back, within their technical lifetime, for the reasons we have considered in this book. These measures usually include: external wall insulation to EnEV standards; roof insulation that requires remodeling of the roof or replacement of the tiles; the replacement of serviceable windows with new, more energy-efficient models; and the replacement of boilers that are still serviceable. In some cases, solar water heating might also fit this category, depending on the building's orientation to the sun, and whether the boiler was due for replacement.

Although these measures are not likely to pay back when real, measured consumption pre-and post- retrofits is taken into account (see Chap. 6), and are very economically inefficient in terms of kWh and CO_2 saved per euro invested, they can have an important place. They reduce heating fuel consumption and CO_2 emissions, and can make homes more comfortable to live in.

The German owner occupiers we have interviewed who have adopted these measures have not told us they did so to save money (Galvin 2011). Their main motivations have been to reduce their personal CO_2 emissions and increase their level of thermal comfort in the home: to save the environment and keep warm. These have been people with above average incomes. Further, the housing providers we have spoken to who have upgraded their properties with such measures (including local authorities) have frequently also expressed the desire to reduce CO_2 emissions and provide warm accommodation for their tenants, with low heating costs.

We call these 'top-end' thermal upgrade measures for several reasons. They are at the top end of energy saving where the marginal costs are high and the marginal returns low. They include the top end of current thermal technology, such as triple-glazed, low U-value windows and highly efficient boilers. They tend to appeal to the top end of the market, where homeowners can afford to invest in their properties to make them more suitable to their needs. In some ways, they have the same appeal as renovating a bathroom or a kitchen: homeowners generally do not invest in these projects to save money, but to lift the quality of their dwelling.

We propose that the government should continue to promote 'top-end' thermal upgrade measures, as it does now—but not on the basis of their being economically viable as this is misleading. There are other good reasons to promote them: they help to protect the environment and mitigate climate change; they can make a home more comfortable to live in; they can make a house look more attractive; they bring thermal technology into the home; they can reduce heating fuel bills, even if not enough to pay back over their technical lifetime; they provide a kind of insurance against sudden, unexpected spikes in heating fuel prices, making household budgeting more stable; they can give a new lease of life to an old building.

However, retrofitting to this level should not be compulsory, especially not for households that cannot afford it. It is seldom economically viable in the commonly understood meaning of the term; it is often an economically inefficient way to save fuel and reduce CO_2 emissions; and as a compulsory standard it makes modest, economically optimum retrofit measures illegal and it puts thermal upgrades out of the financial reach of most homeowners.

We also suggest there is little justification for continuing to subsidize retrofits that go beyond the EnEV standards. As well as 'free-rider' effects that have been observed in several countries (Chap. 3), retrofitting to these standards is even more economically inefficient than to EnEV standards, so that public money is being spent for a fraction of the return it could get in other CO_2 abatement projects.

Government strategy of encouraging 'top-end' measures should adopt a market-led approach and try new, community-based local outreach strategies to address the barriers unique to each community—an approach that has offered clear advantages for example in the 'Neighbour to Neighbour' program in Connecticut US (see Gillich and Sunikka-Blank 2013). Further, since contractors are likely to play a critical role in the success of the policy, not only as builders but also as practical advisors to households, the outreach strategy should be formulated in consultation with contractors, while trying to ensure that the inclusion of the energy efficiency measures targeted by the program offers incentives in the contractors' business models.

We suggest that promoting thermal retrofits at the top-end level needs to be the third branch of heating fuel savings policy, but only alongside the other two. The policy infrastructure for this is already in place, as it represents the main thrust of current policy. But its message needs to be altered, to focus on a particular audience with financial resources, and to be realistic about the economic benefits of this level of retrofitting, based on the actual energy consumption.

9.3.3.1 Saving Potential from 'T' Stream

We have argued in this book that about 1% of dwellings are benefiting from these type of measures annually, achieving an average real, measured saving of 25% per dwelling retrofitted. If the government were to target better-off homeowners with the challenge of improving their homes and protecting the environment at their own personal cost, it is conceivable that this kind of more targeted campaign could increase the current 1% refurbishment rate at this depth slightly to 1.5%. A 1.5% annual retrofit rate at this depth would provide national savings of 0.15% in addition to those provided by economically optimum upgrade measures. This could amount to additional national savings of around 6% over 38 years.

9.3.3.2 Total Saving Potential from CUT Streams

With 15% savings from the 'C' stream, 13% from the 'U' stream and 6% from the 'T' stream this would give total savings of 34% by 2050. Admittedly, this is still short of the 80% goal, but it is far higher than the 9.5% we can hope for from the current policy. In any case, the figures suggested above are based on deliberately conservative estimates, and with a determined and properly informed policy approach they could be even higher.

9.4 Conclusions

In this final chapter, we have acknowledged the successes of the German project of thermal retrofits of existing homes, but highlighted its shortcomings. Its successes are the technical achievements of retrofitting a wide range of homes to high thermal standards; the development of a nationwide infrastructure to do this; and the motivating of a steady stream of well-financed individuals and housing providers to retrofit to these standards. Its main shortfall is its rigid demand for top-end thermal upgrades to the exclusion of other types of upgrades. This leads to economic inefficiency, and prevents homeowners undertaking modest, affordable measures that save far more energy per euro invested. It has also failed to engage vigorously with developments in household behavior change, which are already saving significant amounts of energy and CO_2, and can offer much untapped savings potential.

We have recommended a decisive shift in policy. The focus of attempts to save heating energy in the housing stock should be broadened to a balanced, three-prong CUT policy strategy. The 'C' stream would promote and facilitate modest, cost-effective thermal upgrade measures at an economically optimum level and increase the renovation rate. The 'U' stream would first research and then promote user-behavior strategies for reducing heating consumption without loss of thermal comfort. The 'T' stream would promote top-end retrofits to and beyond current EnEV standards in those households that can afford it and are willing to renovate—not on the basis that these are economically viable, but that they make a contribution to saving energy and CO_2, and can make a home more comfortable and provide insurance against spikes in fuel prices.

Based on our analysis presented in this book, we argue that the current German policy is leading to reductions in domestic heating fuel consumption of about 0.25% per year, whereas this would need to be about 2.1% to reach the goal of 80% reductions by 2050. If policy of the CUT model were to be vigorously pursued, we suggest it is not unreasonable to hope for savings of 34% by 2050, as it could lift the current 0.25% annual consumption reduction rate to well over 1%, without any significant increase in government spending. Further, we suggest that such a policy architecture might work well in other countries of comparable climate and economy that are also concerned to make deeper and cost-effective cuts in energy consumption and CO_2 emissions from home heating.

References

BMWi/BMU (Bundesministerium für Wirtschaft und Technologie/Bundesministerium für Umwelt, Naturschutz und Reaktorschutz) (2010) Energiekonzept für eine umweltschonende, zuverlässige und bezahlbare Energieversorgung. BMWi/BMU, Berlin

Boermans T, Bettgenhäuser K, Hermelink Schimschar S (2011) Cost optimal building performance requirements: calculation methodology for reporting on national energy performance requirements on the basis of cost optimality within the framework of the EPBD. European Council for an Energy Efficient Economy, Stockholm

Faber J, Schroten A (2012) Behavioural climate change mitigation options and their appropriate inclusion in quantitative longer term policy scenarios—main report. CE Delft, Delft. http://ec.europa.eu/clima/policies/roadmap/docs/main_report_en.pdf. Accessed 19 Dec 2012

Flasbart J (2009) Energy efficient residential housing: chances and challenges. In: Presentation at the UNECE conference energy efficiency in housing, Vienna, 23 Nov 2009

Galvin R (2011) Discourse and materiality in environmental policy: the Case of German Federal Policy on Thermal Renovation of Existing Homes. PhD thesis, University of East Anglia. http://justsolutions.eu/Resources/PhDGalvinFinal.pdf. Accessed 19 Dec 2012

Galvin R (2013) Impediments to energy-efficient ventilation of German dwellings: a case study in Aachen. Energy Build 56:32–40. doi:10.1016/j.enbuild.2012.10.020

GdW (2010) Energieeffizientes Bauen und Modernisieren: Gesetzliche Grundlagen; EnEV 2009; Wirtschaftlichkeit, GdW Arbeitshilfe 64. GdW Bundesverband deutscher Wohnungs- und Immobilienunternehmen e.V., Berlin

Gillich A, Sunikka-Blank M (2013) Barriers to domestic energy efficiency—an evaluation of retrofit policies and market transformation strategies. In: Proceedings of the ECEEE 2013 summer study, Presqu'île de Giens, Toulon/Hyères, 3–8 June 2013. European Council for Energy Efficient Economy, Stockholm

Gram-Hanssen K (2010) Residential heat comfort practices: understanding users. Build Res Inform 38(2):175–186

Hargreaves T, Nye M, Burgess J (2010) Making energy visible: a qualitative field study of how householders interact with feedback from smart energy monitors. Energy Policy 38:6111–6119

Schumacher F (1973) Small is beautiful: economics as if people mattered. Blond & Briggs, London

Shove E (2010) Beyond the ABC: climate change policy and theories of social change. Environ Plan A 42(6):1273–1285

Simons H (2012) Energetische Sanierung von Ein- und Zweifamilienhäusern Energetischer Zustand, Sanierungsfortschritte und politische Instrumente, Bericht im Auftrag des Verbandes der Privaten Bausparkassen e.V. Empirica, Berlin

Tschimpke O, Seefeldt F, Thamling N, Kemmler A, Claasen T, Gassner H, Neusüss P, Lind E (2011) Anforderungen an einen Sanierungsfahrplan: Auf dem Weg zu einem klimaneutralen Gebäudebestand bis 2050. NABU/prognos, Berlin. http://www.nabu.de/sanierungsfahrplan/NABU-Sanierungsfahrplan_endg.pdf. Accessed 21 Jan 2012

UBA (Umwelt Bundesamt) (2006) Wie private Haushalte die Umwelt nutzen—höherer Energieverbrauch trotz Effizienzsteigerungen. UBA, Dessau

Printed in the United States
By Bookmasters